Bernie Paul

WIND ON MY WINGS

In loving dedication to my wife BEHRI

WIND ON MY WINGS

BY PERCY KNAUTH

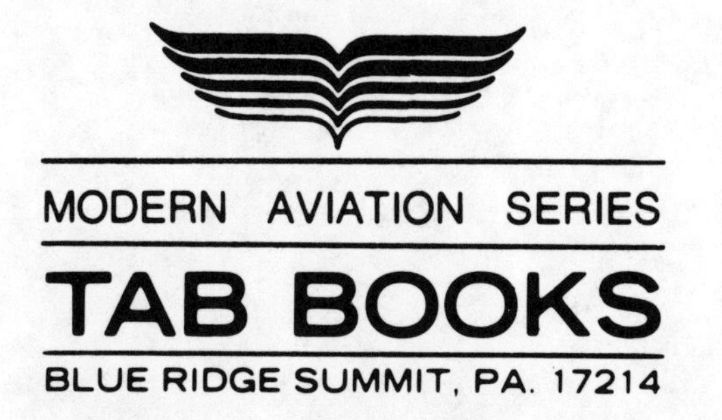

MODERN AVIATION SERIES

TAB BOOKS

BLUE RIDGE SUMMIT, PA. 17214

FIRST EDITION

FIRST PRINTING—FEBRUARY 1980

Printed in the United States of America

Library of Congress Cataloging in Publication Data

Knauth, Percy, 1914- Wind on my wings.

First published in 1960.
Includes indexes.
1. Knauth, Percy 1914- 2. Air pilots-United States—Biography. I. Title.
TL540.K56A33 1980 629.13'092'4 (B) 79-25225
ISBN 0-8306-9968-6
ISBN 0-8306-2278-0 pbk.

Photo of Bonanza V35B, courtesy of Beech Aircraft Corporation.

Foreword with Hindsight

When I embarked upon the project of this book, the first thing that I did was to write a Foreword. This, it seemed to me, was the straightforward and logical thing to do—a Foreword introduces the writer as well as his readers to the subject matter of his story, gives him a chance to editorialize a bit about it and, almost always last but never least, a graceful opportunity to thank all those who helped him with the book. And so I visualized my task, and sat myself down to do it, and finished a Foreword that ran to about fourteen pages.

Alas, today those words, all but a few of them, are in the wastebasket. It isn't that they turned out to be trash; they didn't, or anyway not all of them. The statistics on private flying, for example, still make an impressive point: 10,000,000 hours per year logged by people flying for business or pleasure or both in their own or their company-owned airplanes; more than three times the total time flown by scheduled commercial airlines. And the fact that the number of pilots in general aviation, which excludes the airlines, is now well above 200,000 with thousands more learning to fly every year. And that a safe, modern small airplane today can fly anywhere in the world, over continents and over oceans (as I was to learn for myself), utilizing all of the navigational aids and automatic pilot devices available to the airlines; and this for as little as ten cents a mile, about the same cost as an automobile. All these things are still as true as when I first wrote them in my original Foreword, except that the performance figures are going up and the cost figures down as more and better airplanes and more people get into the flying game.

What happened was that the experience of flying overwhelmed me. For thirty-odd years of my life—since before Lindbergh flew to Paris—I have wanted, with waxing and waning passion, to learn to fly. I thought I knew pretty well what it would be like; I had flown a good deal in airliners and more than the average person in small planes. So when at last

I decided the time had come when I could give it an all-or-nothing try, I approached it as a man might approach marriage with a girl he has known for many years—with hope, with expectations, and with pleasure that at long last the dream is going to come true; but with the knowledge, too, that this was a world not entirely unknown to me, that mostly it was a matter, so to speak, of taking the vows and settling down to enjoy permanently a relationship whose delights I had tasted intermittently over the years.

How wrong I was! Marriage is never like that, and neither is flying. I discovered that, far from entering into the fascinating world of the air, I had in all my previous experiences only approached its threshold. The difference between being flown by someone else and flying yourself is the difference between reading a story and writing one, or studying a painting and painting one, or listening to a song and singing. The first time a fledgling pilot takes to the air he is reborn; he sees things differently, feels things differently, for he is a man who has tasted a freedom beyond compare. And he has known a challenge, too; the kind of supreme challenge which life today does not present very often to a man. That is why no flier, no matter how far or how long he has flown thereafter, has ever forgotten the magic quality of that first flight, his solo.

I also found in flying a different kind of person. People who fly have a different outlook on life and the world from people who don't. Not that they look different, or dress differently, or even talk differently; nowadays a businessman can go from his office straight to the airport, get into his airplane and fly six hundred or seven hundred miles without taking off his hat. And at the business conference or dinner party or whatever it is he is flying to he probably will not even mention this flight, which a bare twenty-five years ago would have meant wearing leather jacket and helmet and goggles and risking his neck every minute of the way.

No, he probably wouldn't mention it—except to another flier. But let one flier find another, or even just somebody who is interested in flying, and they will talk for hours. They will re-create all the things seen and felt in that wonderful world of air: the sense of remoteness from the busy world below, the feeling of intense brotherhood formed with those who man the radio ranges and control towers and weather stations that bring the pilot home, the clouds and the colors, the surge of the wind on their wings. They will speak of things that are spiritual and beautiful and of things that are practical and utilitarian; they will mix up angels and engines, sunsets and spark plugs, fraternity and frequencies in one all-encompassing comradeship of interests that makes for the best and most lasting kind of friendship any man can have. And they will speak of time and distance in a manner to make the neophyte's head swim.

I remember listening to one such conversation shortly before I started learning to fly—a pilot describing his day. That morning, after breakfasting at a comfortable hour, he had taken off with his wife from Teterboro, across the Hudson from New York. They had flown out to Montauk Point, about 150 miles away on the tip of Long Island, for a swim and a visit with some friends. They had taken off from there and landed for lunch on Block Island. After lunch, the pilot had flown to a business conference in Hartford. He had left that in time to spend a half-hour circling over the America's Cup races off Newport. He landed at Martha's Vineyard in time for a swim, cocktails and dinner before taking off to fly back to Teterboro where he landed, after a beautiful night flight, in time to get comfortably to bed.

Time expands and distance shrinks for the flier. It doesn't take long for even the neophyte to learn to think of a hundred miles in terms of less than an hour of relaxed, straightline distance as he sits in his airplane watching the earth drift by. Soon he finds himself thinking about spending a week end at someplace hiterto totally unreachable in that time by car. In just one day, not long ago, I flew from New York to Texas with my friend Bill Strohmeier, of the Piper Aircraft company, landing there in plenty of time to see various friends, have a swim in our motel pool and an early dinner. Three days later, having meanwhile flown on to San Francisco, the United States of America were reduced for me to sixteen hours of incomparably beautiful flying with three stops for gas at Rock Springs, Wyoming, Omaha, Nebraska, and Cleveland, Ohio; we left San Francisco at 8.30 A.M. Pacific Time and landed at La Guardia Field in New York at six in the morning, Eastern Daylight Saving Time.

These were experiences that no books, no moves, no dreams could ever remotely convey. I had to live them to realize what flying really was. And that is why this book has turned out so differently from its original concept. My facts and figures, the carefully compiled data with which I intended to bolster my own statements about what modern flying is and what it can do for those who take it up, went into the discard. Instead, there came to be on paper my own intensely personal story about my discovery of what flying is. Most of this book was written immediately after making certain flights which were particularly meaningful in that discovery, for I wanted to get my feelings into words while they were still fresh within me, to communicate as quickly as I could my own excitement of these early ventures into the reality of the air.

I envy those who still have the experience of learning to fly ahead of them. I hope they will find everything in it that I found, and more. I hope that they will be as fortunate as I was in the people who help them on their way. I had Steve Gentle, at Katama Airpark on Martha's Vineyard. He is the greatest teacher I have ever known. After hundreds of students and thousands of hours of instruction he still gets a boy's fun out of flying, and he communicates it to his students every step of the way. I also had Russ Swanson and Stan Kenecko, at the Danbury School of Aeronautics in Danbury, Connecticut. They taught me the fundamentals of cross-country flying, how to read maps and how to read the ground so that if radio aids fail I can still find my way home. There have been times when I have blessed them for their insistence on my learning pilotage, which is what flying by map is called; and there will be more because even the best of man-made systems sometimes have to be retired for overhaul so that a radio range which should be working isn't.

There were, of course (and will be for every student pilot always), those who sweated out this project on the ground. My friend Richard W. Johnston, assistant managing editor of *Sports Illustrated,* where I earn my living, is still convinced that the Great Lakes trainer, of acrobatic fame in the thirties, was the last great airplane built and he regarded my ventures aloft in a modern Tri-Pacer with serious misgivings. I am pleased and proud that he and Sidney L. James, then *Sports Illustrated*'s managing editor, thought well enough of what I have written here to publish parts of the book in their magazine. My sons Peter, Philip, Alan, and Stephen all went flying with me at one time or another, giving me the gift of their confidence, which I cherish.

What about your wife? What does she do while you're flying? People are always asking me that question. My answer is: she was great. She let me go, and waited. And she was my first passenger when I got my private license. Now she wants to learn to fly herself. She says it's because she wants to be able to land the plane if anything should ever happen to me while we're aloft, but I know differently. I think she lost her heart to flying on a golden day when we flew over to Nantucket and saw the island and Cape Cod spread out below us like a painting framed by the deep blue of the sea, and on another day when we came slanting down through the haze of an Indian summer afternoon to gently touch our wheels on the runway at Teterboro, one of the busiest airports in the world. She loves to fly, and she says it was worth all the waiting.

One more picture comes to mind, one of my early impressions, one that convinced me that now was the time for me to learn to fly myself. It was a cloudy day, and Bill Strohmeier and I were flying home from Martha's Vineyard. Under the overcast all the world was silvery and subdued; a silvery light came filtering through the gray ceiling above; lakes, rivers, ponds were silver splashes in the gray monochrome of the earth below. The air was very still. We flew toward a curtain of rain in the far distance, smoothly, steadily, our plane hanging between earth and sky on the invisible buoyancy of the quiet air. We were utterly alone, but the voice of our brothers on the ground came reassuringly and musically clear in the beeping signals of the radio range by which we navigated. The hands of the clock crept slowly forward and now we approached our home field. Quietly Bill spoke into his microphone, announcing himself to the airport tower. We turned and entered the traffic pattern and, as the runway swung into view ahead, started our final glide. The engine died away to a murmur, the wind sighed gently past our wings. The ground came slowly toward us; now it was very near; now the runway flowed swiftly past below; our tires felt the pavement; we were down. Lights blinked in the dusk, the tower cut into the silence, the airplane turned and taxied toward the hangar. We were home.

Of many, many pictures of the new life I found in the sky, this remains one of the most beautiful. Perhaps it is because this flight was an introduction to me of the world that awaited me; in any event, I still turn back to it, and to the friendship of Bill Strohmeier, with whom I flew that day. Without his help, and that of the Piper Aircraft Corporation, I never would have learned to fly at all.

New York, N.Y.

Contents

WIND ON MY WINGS

1. *The Runway*

From the crest of a long hill that skirts the southern edge of Danbury a bumpy country road winds sharply downward and there, on a curve, is a point where the entire airport lies spread out to view, almost as one would see it from a landing plane.

Route 7 with its traffic and its telephone wires borders the near end; beyond, to the left, are the small, square buildings and parked planes of Cliff Sadler's aircraft service, facing, across the wide field, the stands and stadium of the Danbury Fair Grounds. Far down near the other end is the squat hangar of the Danbury School of Aeronautics, where I fly.

Each day, on my way to the field, I slow down at this curve to get a preview, so to speak, of the moments in the plane to come. Here, on the ground, I can anticipate the feeling of the air, the little tensions, anxieties, and exhilarations of that last long glide from the air world to the land world. This is the time when the two begin to meet, when the qualities of the one become a part of the problems of the other. The highway and its wires—we will sweep over them, not too high and certainly not too low. Cliff Sadler's place will merit a glance to see if airplanes are taxiing out from it. I mark the little red-and-yellow Aeronca training planes beetling along the grass—they will be in the traffic pattern when, ten or fifteen minutes hence, I will be in the air. I check the wind tee to see whether I will have a cross wind, and try to find some smoke somewhere from fire or chimney that will tell me how strongly the wind is blowing. And finally, I gaze down the runway, that broad, gray, tire-streaked expanse of truncated boulevard that in these days and weeks of learning has become so much a part of me that I even see it in my dreams.

To the fledgling pilot, the runway is a symbol that appears in many changing forms. It is home port, safe haven, mother's arms, a wife's warm breast in the communal comfort of bed after a long, hard day. It is also the light beckoning to the mariner, warning of shoals but promising land. The experience of flight is a very concentrated one; it has many of the elements of voyaging on the sea, but it packs into a few brief moments what days or weeks on the ocean might contain. What happens in the air happens quickly, and this is nowhere more true than in those moments of landing when the pilot, earthward bound on his invisible but usually lively channel of air, reacts with hands, feet, eyes, and ears to the many different movements, sights, and sounds that characterize that border area between earth and sky.

Thus, in the early period of learning to fly, the runway becomes a symbol of ultimate achievement as well as of safety and rest after labors aloft. To approach just right, riding the wind as surely as a ship sailing smoothly into harbor, is to know one of the great joys of flying, the first sense of mastery over the new and intangible element of the air. To flare out at just the proper instant, floating a foot or so above the runway until the wheels touch lightly down, is a reward to all of the senses so actively employed in bringing the plane home. There is no feeling I know to compare with it except returning from the sea: it combines the satisfaction of having applied a new skill well, of having met a challenge successfully, and of having made a safe return from a world full of strange fascinations to the world one knows, all in one.

To me, at the age of forty-four, the runway is also symbolic of a dream long pursued and never fulfilled. As a boy, the dream of flying filled my waking and sleeping hours. Sitting on the floor of our living room, inside two chairs which I laid on their backs to simulate a cockpit, I could think myself into the air for hours on end. Always it was the moment of coming back to earth which I savored most: I could feel the perfect three-point landing to which I sailed down out of sunlit skies, moving my feet on the imaginary rudder bar, my hands guiding the sawed-off broomstick which controlled imaginary ailerons and elevator on imaginary wings and tail. And when I learned to paddle a canoe, it was not a canoe that I brought into the dock but an airplane, paddling hard as I approached head on, then cutting my power to glide in, steering gently with the paddle, gradually turning parallel until, at the precisely proper instant, I swung broadside on and, all forward motion stopped (my plane in a full stall), lightly kissed the land.

The dream never lost its power, but with the years the gap between it and reality widened. I did not realize this until I started, at last, to fly. The boy would, I am sure, have found the world of air, despite all his imaginary excursions into it, as new, as strange, as exciting as I did. The man found all this—but he found, too, that the years had brought other factors to intervene between him and the dream: an inescapable sense of responsibility for wife, children, career which had to be fitted into the pros and cons, a natural sense of caution which had matured into acute awareness of risk, a feeling at times that he either should have done this a long time ago or should never have tried it at all.

The risk of flying even the smallest private planes today is minimal; aircraft and engines have been developed to so high a degree of dependability that their failure in the air is virtually unheard of. The pilot who takes proper care of his airplane need never doubt its loyalty; but he should doubt himself. Like the sea, the air is intolerant of carelessness and stupidity; and to a man approaching his middle years, accustomed to the easy exercising of the basic skills of living, his early fumblings in a strange machine and an apparently unstable element can make him feel unsure.

And thus the runway also appears to me as a symbol of doubt of my own self. Swimming around there in the broad vista before my eyes, distant focal point of my fears and my desires in the all-encompassing panorama of earth and sky, it both attracts and repels me, yet lures me always on and down.

I first flew from the runway of the airpark at Katama on Martha's Vineyard, close by the old whaling port of Edgartown. The runway here is grass, and it has been dispatching and receiving airplanes for some thirty-odd years since the day when a Curtis Robin first landed on it with a crew of air-borne picknickers sometime in the golden twenties. Long usage has given the field a worn and friendly look, like that of land which has been tilled for many generations; but its modern function is clearly shown by the runway numbers, indicating the points of the compass toward which each heads, that have been carefully cut into the springy island turf and etched out with bright white sand from the beach nearby. Katama has always been, and still is, primarily a landing field for small airplanes, but airliner-sized craft can and have landed and taken off there too.

The airplane was a Piper Tri-Pacer, four-place, shiny red and white, brand-new, with the number N-9013-D (for Delta) painted on its sides. One-three-Delta will always have a very special place in my heart. As a vehicle to aspirations which were cloudy both in the figurative and the literal sense of the word it was perfect: if ever an airplane was built which could lure a somewhat self-doubting Thomas into the air and keep him there, the Tri-Pacer is the one. Its 160-horsepower engine has more than enough power to overcome the initial fumblings of the neophyte flier. It has a tricycle landing gear on which it sits level on the ground, giving the student the familiar feeling of being in a car instead of the immediately strange and unsettling, tipped-backward sensation of the now outmoded tailwheel gear. Its control wheel and rudder pedals are linked by springs so that the use of either one will generally steer the plane. Finally, it is completely and reassuringly comfortable, with a wide, curved windshield in front, large windows shaded by the broad wing, and seats cushioned with foam rubber, trimmed in happy colors, and adjustable frontward and backward to assure maximum ease in reaching and handling the controls.

In addition, the Tri-Pacer is a remarkably stable airplane which, once trimmed out for level flight at cruising altitude, will practically fly itself, requiring only the lightest occasional touch to correct for wind draft or the fleeting effect of bumpy air. I have seen One-three-Delta trimmed out so delicately, in fact, that by nodding his head forward or backward the pilot could cause the nose to dip or rise. Like all modern private airplanes, the Tri-Pacer's natural tendency is always to fly straight; it is inherently stable, will resist efforts to get it into a tailspin and if forced into one, will come out into a straightforward dive after one and a half turns if the pilot simply lets go of all controls and allows the plane to take care of itself.

All these things I had been told, before I first stood on the runway there at Katama and looked at this winged sedan in which I was to take to the skies, but I had yet to appreciate what they would mean to me, the student pilot, when I first ventured aloft. It is one thing to talk about a revolution in private flying and quite another to experience it; but a revolution there had been, in the last decade, as the Tri-Pacer is here to testify.

Now I climbed into the plane and looked at the world as I would henceforth see it—the earth-sky world of the flier. One ends, the other begins, and the runway is the springboard from one to the other; and the sky is never again just the empty sky, but a place of winds and clouds and currents, as mysterious and as fascinating as the sea. We run through the little cockpit chores that are a necessary and anticipatory prelude to every flight; start engine, check oil pressure immediately (if it doesn't come up in 30 seconds shut off the engine; something is wrong), check fuel gauges, check which tank we're on, move all controls to be sure they are working freely. Now we taxi to the runway (I steer with my feet and find the airplane answers as smoothly and easily as my pushmobile of childhood days). We reach the end of the runway, turn into the wind and set the brakes. A quick check of the oil pressure and oil temperature gauges; then I advance the throttle, pushing in the big red knob until the engine is running at 1,800 r.p.m. We check the magnetos, switching from both first to the right then back to both, then to the left and back to both again, turning the key in the ignition lock (if the r.p.m.s drop more than 100, a magneto is not delivering a proper spark and plug or magneto should be checked). I check the carburetor heat, pulling out the knob that turns it on; when it is applied and working the engine slows appreciably. We check the altimeter for proper setting, check the trimming handle set in the ceiling above to see that we are trimmed for take-off. Finally, I reach down between the seats to where a lever sticks out like the hand-brake on a sports car and pull it up one notch to set the flaps in first position. Just beyond my window I see the flap come down, a section of the trailing edge of the wing, a simple but wonderful device that considerably increases the lift of the wing at slow speeds.

Now I release the brake, we turn in a wide circle and scan the sky for planes that might be coming in to land. Sky and runway are clear; we trundle out and line up for the take-off.

Sitting in the airplane that first day, looking the length of Katama's worn brown strip, life reached a corner where I stood for an instant, irresolute, not yet compelled to turn. To push the throttle forward was an act like pushing forth from the safety of land for a voyage of unknown duration on uncharted seas—an act at once immensely alluring, challenging, exhilarating yet touched with desperation too, and irrevocable for me. The engine seemed to shout into my ears; then it pulled, we moved, we rushed ahead, bumped briefly on the earth, and soared.

The runway, in the days that followed, became the central fact of my existence. On the ground, it stretched before me like a clipped grass avenue, by turns inviting, by turns a grim challenge, leading off and away toward the beach and the open sea. My feelings toward it fluctuated with my nascent flying skill, but even when I hated it for a perverse arena in which all my shortcomings were mercilessly exposed, I never lost the background feeling of exhilaration that came with this effortless leap into the air. Down the open, unencumbered springboard of turf the Tri-Pacer moved as smoothly and assuredly as any automobile; at around 60 miles an hour it simply moved from the ground into the air. A third of the way along the runway there was a slight depression; at this point I usually became air-borne, the grass falling away beneath my wheels, the plane lifting free. And the runway, a brown blur, grew smaller, turned into a dune, a beach, and then I was climbing over the marbled sea.

By the time I had climbed out, turned left in the standard rectangular pattern and headed on a tangential course toward the beach again, everything was different. At 800 to 1,000 feet all ties to the earth were already severed and the plane floated alone, remote, a creature in its element. Motion ceased; instead of the headlong plunge of the take-off and initial climb we seemed to drift now, almost languidly, over blue sea, white beach, deep green of the alfalfa field bordering the airport. The engine, throttled back to cruising speed, had lost its urgency (I later found that its steady drone would grow so familiar that I stopped hearing it entirely; only a miss in the regular beat, a sudden silence, would be deafening). I rode in a world of purity and clarity that tuned and challenged all my senses; I felt the surge of unseen currents, a rising and falling of irresistible and invisible waves, and gradually there came to me a feeling of enormous, gravityless power in which a tiny movement of my hand or foot could cause the earth to wheel or rise while the airplane, as though hung on gimbals, seemed to remain steady in the sky.

And then there came the feeling of aloneness. Up here, no sound from earth could reach me unless I chose to let it—no man's cry, no intrusive voice commanding haste lest time be wasted, trains be missed, work be left undone. I could speak to the world by radio; but the world, if I decided to ignore it, could not speak to me. And on the best of days, when the plane droned along on air as smooth as limpid water, there was peace of a kind I had never felt before—the peace of utter solitude, when life and the world fade into the misty distance of infinity and the infinite becomes tangible through the communication of the soul. The sky took on a grandeur then; the little plane was touched with a celestial magic; it was no longer a mechanical contrivance that stayed aloft in accordance with well-known aerodynamic laws but the creation of unearthly hands, a vehicle born of dreams in which I was privileged to enter into an unpeopled sphere reserved for me alone.

But life is not all magic, even in the sky. There is work to be done aloft, and it centers ultimately on the runway, the beginning and end of every air-borne journey. And as free, as untrammeled, as limitless as the sky may be, it too, around the runway, must be circumscribed by a man-established pattern; and that pattern, an inviolate discipline for every airplane that lands or takes off from any airport anywhere, in the weeks that followed became the pattern of my flying hours.

This is how it goes: take off, climb to 500 feet above the runway, turn left 90 degrees, climb to 800 feet, turn left 90 degrees and cruise parallel to the runway on the downwind leg, turn left 90 degrees onto the base leg, slow to 100 miles per hour, apply first flaps, throttle back to idling speed, turn left 90 degrees heading into the runway on the final approach, apply full flaps, come down and land. Invariable and inviolate except by special instruction, this pattern is universally established so that the increasing number of aircraft in the sky will always know what to expect of each other when leaving or approaching the ground.

In the pattern, time seems crowded to the fledgling flier. A Tri-Pacer at full throttle can climb at nearly 1,000 feet per minute; even with reduced power (there is no use in flogging an engine unnecessarily and with 160 horsepower there is plenty in reserve), I would get up to 500 feet, the point of my first turn, in not much more than half a minute. In that short time I had to establish a good climbing attitude (90 miles per hour at 2,350 r.p.m. is reasonable), start correcting for possible wind drift, throttle back, take off flaps (slowly, very slowly I released that lever, meanwhile inching the wheel back to compensate for the loss of lift), look around for other aircraft and start my first, climbing turn. By the time I was out of that turn I was at 800 feet, the pattern height, or better; I had now to level off, trim ship to cruise attitude, throttle back to cruising speed (115 miles per hour, 2,150 r.p.m.), check my position in regard to the runway, check the sky again for other aircraft, and start my second turn, into the downwind leg.

Now came a moment of brief respite. All things being reasonably well done. I was proceeding straight and level, parallel to the runway, with the landmarks that counted in plain sight and all instruments behaving. I could uncurl fingers and toes from wheel and rudder pedals and relax. I might, if the wind was off the runway, find myself drifting a bit and turn slightly, crabbing in toward it. It might be jouncy up there; I learned the quick reflex movements which bring up a tipped wing, correct for a sideways swerve. And then it was time for base leg and final.

Here, 800 feet high and still an improbable distance from the runway, is where a landing begins. It is a pattern of actions and movements, reasonably timed, reasonably precise, beautifully logical, harmonious with the plane, the air, and the distance; and if all of its parts add up to a perfect whole, the aircraft will touch the ground at the instant it loses its flying speed and at the spot where the pilot wants to be. Like bringing a canoe to the dock, or coasting a car down the street, into the driveway and precisely into the garage, it is a matter of perception, judgment, and practice and if the end result is good, it is one of the most satisfying things in the world.

Turning into base leg, I would pull on the carburetor heat, the first step in cutting the power. The act became automatic after a while, which is as it should be, for carburetor heat is a vital adjunct to an aircraft engine: it prevents ice from forming in the carburetor jets and intake manifold, which might cause the engine to choke and die if the throttle is suddenly advanced. With the heat on, the engine slowed; I could now throttle back and, at 100 mph, put on first flaps. Off to my left, the runway gradually swam into view. It was time to start closing the throttle. As the engine noise died away, the nose dropped and I could hear the wind whistle past struts and wings outside. Now came the turn into the final approach, the ground pivoting below. I eased on full flaps and lined up with the runway. We were coming down fast now. Pushing the wheel against the lifting, slowing force of the flaps I held the air speed to 85 miles an hour, pointing toward the field. Below, the green alfalfa flowed past in a swift blur; I saw it only as a sort of backdrop of color; my eyes were fixed on the runway ahead.

This was the picture, these were the sensations that I carried right into my dreams. Close to the ground, the air caught and pulled at my wings; often it seemed that the plane itself did not want to land. The ground seemed strange and foreign, an alien element onto which I was forcing myself out of the friendly sky. Where I was floating before, I seemed to be rushing now with geometrically increasing speed. It is a trick which the runway always plays on the neophyte: by instinct he stares straight ahead, over the nose of the plane, eyes fixed on the ground which comes up at a sharp angle. It is difficult to realize, because it seems contrary to reason, that by looking off to one side, not at the runway but far down it, things can be made to slow down and assume their proper perspective. Then the runway will gradually flatten out, tilting, as it were, to meet the plane which at the proper instant, by easing the wheel completely back, can be flared out a foot or so high until it loses flying speed and touches down.

This is that intangible sixth sense which fliers call "the feel of the ground." It is a descriptive phrase, and well chosen, for it is indeed the re-establishment of a relationship from one element to the other, the bridging of a gap between the earth and the sky. It cannot be studied, it cannot be taught, it can only be acquired; and the day it is acquired is a red-letter day, one that calls for celebration and song. For suddenly everything seems to come into focus; the relationship is miraculously there; the runway is no longer repelling but an invitation to excellence, and the student knows he need never really fear landings again. The bridge has been struck between the air and the ground, and it will stand solidly forever. There will be many a bumpy landing yet as the student learns to polish his skill, but there will always be the fine, firm conviction that, bumps or no bumps, he can get the plane home.

It was at Katama that I experienced this feeling for the first time; and I knew, even at that very moment, that there would never be another flight like this for me, anywhere or anytime. It was a gray, cool day when Steve Gentle, my instructor, waved me out alone for the first time. Three times he had ridden me around the pattern on this morning, driving me hard, scolding me vehemently for the least mistake. Now I knew why—now it was up to me to go around, all on my own.

It was an appalling, inspiring, unforgettable instant: the short-lived sense of panic, the urgent desire to be somewhere else, far, far away, the sudden determination, the irrevocable act of pushing the throttle forward and taking to the air. I can still see Steve's stocky figure, tiny on the runway, as I rushed past, climbing, soaring. I can still feel that first landing and the inexpressible sense of joy it brought me—a joy that time does not diminish, a sense of accomplishment such as I had never known.

At Danbury, other problems face me now—new problems of terrain and wind and air and navigation that go through my mind as I drive down the hill to the field. I am reaching higher, exploring the wide, wide world of sky above the pattern where clouds sail and the wind blows free. But the runway, be it here at Danbury or at Katama or anywhere else where I may someday land, remains symbolic to me of a world in true balance, in which the moment is all-important and the individual his own king. Soaring down his channel of air toward his haven, the flier is beyond the reach and aid of his fellow men, a prospect frightening at first but one which, with acquired skill, becomes a matter of wholly satisfying pride. It is the pilot alone who can bring his plane safely down, and no pilot can get by for long with less than a wholehearted effort and a sound application of the skills which he has so diligently learned. For the runway is the final goal of a pattern of true craftsmanship, of skills acquired without shortcuts, without excuses, and learned only one way—well.

Through the weeks and months of learning to fly, this is my greatest reward: to touch down at the end of a flight and know, and understand, as my wheels kiss solid ground again, what it is like to have held my life in my own hands.

2. *The Air*

STUDENT PILOT TALKING TO HIMSELF: *"All right, Ace, this is it now. From here on you're on your own. No more Steve or Russ or Gene to bail you out when you do something wrong; no more anybody but yourself. You're finished with pattern flying now; you're going up into the air, alone."*

It is a beautiful, clear, cold midwinter day. This morning, early, the air was absolutely still; around ten o'clock it began to whisper a little in the trees. Now, at one, as I come down the long hill toward the field I see in various little ways that it has picked up considerably. The laundry hanging behind the little gray house beside the road flaps awkwardly, frozen stiff. Smoke from a brush fire blows in puffs across the runway. The wind tee down there points west, but the wind sock rises and falls jerkily at an angle to it—that means a variable, gusty cross wind. My stomach tightens slightly; it will be rough on take-off and landing.

But wind has to be learned, and I will learn it today.

Gene, the instructor, sends me out to the airplane alone, for the first time. "Take it up to 2,500 feet and practice turns," he says. "Watch your take-off; it's gusty, so you want to be ready to hold your wing down if you get one blowing cross. It'll be windy up there, too. Remember, in your turns, to hold it shallow on the upwind, steeper on the downwind side. Keep the airport in sight, don't get lost, and hold your altitude." He climbs into an Aeronca with another student, cranks up and taxis away.

I walk out to the red Tri-Pacer sitting on the ramp in the sun. My responsibilities lie heavily on me. This is my airplane now for the next hour or so—mine to take into the air, mine to bring safely home again. The world of the air is still an alien place to me; I venture into it with trepidation. First of all, I make sure that everything is all right with my airplane.

I check the oil—it shows over seven quarts on the dipstick, which is fine. Spark plug leads and fuel line connections are all okay too. I run a hand along the propeller, checking for nicks and dents, and give it a shake to make sure it is tight and secure. Then I walk out along the right wing, running my hand along the leading edge as I go—the caressing movement, which so aptly expresses my feeling toward this winged love of mine, will also tell me of dents or cracks in the fabric. At the trailing edge I made a careful check of hingepins and control line fastenings on ailerons and flaps. I check rudder and horizontal stabilizer the same way, then out and back again along the left wing. Finally I drain a bit of gasoline out of the two sediment bulbs below the engine to make sure that my fuel is running sweet and clean, take a look at the tires, check for leaks in the hydraulic brakes, snap the generator belt for tautness, and climb in. The whole routine takes three or four minutes, and adds immeasurably to my peace of mind.

It feels strangely lonesome in the little cabin with that empty seat beside me. I prime the engine and start it; she comes easily and the steady ticking over of those four stout cylinders is reassuring. The oil pressure comes right up; both tanks show full and I am on the right tank now. Controls are free, and I release the brake and taxi slowly out to the end of the runway. There I run my engine up to 1,800 r.p.m., check my magnetos and carburetor heat, put on first flaps, swing around for one last, lingering look at sky and ground, and launch the Tri-Pacer into the blue-gold afternoon.

One of the greatest attractions in flying is the infinite variety of the air. It never ceases to surprise me that each flight can be so different from the last, and each one such a challenge. Outwardly all things may seem the same—there is the old, familiar runway stretching out ahead, the brown hills beyond, the blue sky waiting with a few small, high, wind-raveled clouds. Yet every time there is a difference—a difference noted half-consciously, half-instinctively when the moment comes to push the throttle forward. It is the difference in the air, and in its most tangible manifestation, the wind.

I feel it at once, as my engine roars, my plane begins to move, and the air-speed needle comes off its peg and starts around the dial.

First there is the growing rush of sound, sound which is almost immediately transformed to feeling in my hands on the control wheel. The blades of my propeller disappear, there is not even a blur in front; but the wind, the wind of my passage which will eventually bear me aloft and sustain my flight, becomes a felt and living thing. It rocks the airplane a bit; I hold her steady by pushing the wheel slightly forward to keep her nosewheel firmly on the ground so I can steer her. But now the wind begins to pluck lightly at the wings. I can feel it—tiny, lifting movements as the air-speed needle goes past 50 miles per hour. I am reaching flying speed. Still I hold her down, until we are doing 60 and the runway is a streaking river of gray. Now I relax my forward pressure on the wheel, easing it back gently to see if she is ready. The plucking on the wings grows stronger; lift and gravity are almost equalized. I am beginning to ride on surging waves of air. At 65 I ease the wheel back still more, feel the nosewheel leave the ground, feel the main gear bounce lightly, once, twice, perhaps three times—and then we are flying.

This is the critical instant—the moment of freedom, the moment of complete and utter change. In one small, swift movement, one light leap, I have left the clinging, solid earth and been reborn into another, totally different world. I may be only six inches off the ground, or even three—but now the wind takes over.

I don't really *feel* it, as I would on the ground. It isn't like wind striking the sail of a boat, heeling it over, or wind buffeting me in a car. Here there is no relationship of wind to something planted solidly, or even any feeling of solid pressure in the wind itself. That is a part of the big change from earth to air. On earth, my instinct is to fight the wind—to push the rudder of my boat over and press my surging sail back against the wind's pressure, or hold the wheels of my automobile firmly on the road and resist the gusts that buffet me. I cannot do that here—I have no solid point on which to stand and fight. The wind is a part of the ocean of air in which I ride; to fight it is as impossible as trying to punch air, as useless as roaring out imprecations into a hurricane. What I must do is compromise.

The wind is pushing me sideways; I want to climb straight ahead. I tip a wing, just slightly, into the wind's direction. The airplane banks, a tiny bit, and I apply a touch of opposite rudder to counteract its desire to turn. The effect is almost magical—the runway stops sliding sideways past me, and I am flying straight. The slightly tipped wing has the effect of sliding the airplane just a little in the direction of the wind, conpensating for the drift; the compromise keeps me precisely on a straight line.

I am climbing now at the rate of about six feet every second, reaching for the sky. But the wind is restless. It shifts and changes, blowing in those variable gusts that made me anxious down below; my hands on the wheel are in constant motion, correcting, feeling it out as we climb. The surging billows of air have shortened into real bumps now—up, down, over and back. It is as rough as I thought it would be; it is like climbing up a steep and bumpy hill. The peace and beauty of flying is sometimes hardly won; and this is one of those days.

But I want to make this climb a good one, wind or no, and so I concentrate on the job. At 200 feet I take off flaps, easing the wheel back to compensate for the loss of lift. Throttle back to climb power—2,350 r.p.m.; airspeed 90 mph. She jumps around; it's hard to hold the air speed steady. I reach up to the trim crank and give it a couple of turns toward UP—that puts her in a climbing attitude and lessens my job with the wheel. I look back at the runway, receding behind and below, and around at the sky for other airplanes. At 500 feet we turn left, climb to 800 in the pattern, then start a sweeping, climbing right turn up and out and away.

Despite the work of climbing it is incredibly beautiful up here. Already I can see for miles and miles. Lakes and ponds are crinkled blue in a rolling landscape of brown hills. My cabin heater warms me comfortably; I sit in a snug little world. From horizon to horizon the air is crystal clear. Houses and towns have shrunk to toy proportions; Danbury, with its bustling streets and crowded traffic, is nothing more than a peaceful mosaic to me as I pass over it at 2,000 feet, still climbing.

At 2,500 feet I level off, cut back to cruising power, 2,150 r.p.m., complete my turn and head for Lake Candlewood, a few miles north of the town. The bumps have vanished; the air is smooth and peaceful. That wind below might never have been, so calm it seems up here. For a moment I am almost tempted to think that I have climbed beyond it—but not for long.

Lake Candlewood begins to seem an evanescent goal. Not that it is a small lake and hard to find; on the contrary, it is about twenty miles long and stretches right ahead of me. But I don't seem to be getting to it; though I am pointing right at it I seem to be drifting away from it, off to the left of it, all the time.

Looking down over the side, I suddenly see why. That peaceful countryside is not unreeling straight below me. Once again, I am sliding sideways. The wind is coming almost directly from my right, pushing me hard. I've got to angle over into it if I want to get where I'm going.

I turn the airplane slowly to the right, watching the ground meanwhile. The sideways motion ceases gradually as we turn into the wind, until I have achieved my compromise: now I am flying straight over the ground, as I should be. I look up, out and ahead. My nose is pointing to the right of Lake Candlewood. Who would have thought it! I'm flying like a crab, sideways along my track, but this is the way I should make it—and I do.

For the first time I actually comprehend what I have learned on paper on the ground—the wind triangle. Here is where I am; there is where I want to go; here is the direction and force of the wind: somewhere in between is the compromise course I must fly. Wind and forward speed, balanced in this practical equation, will together bring me there. I look down at the lake, my first destination reached all alone, with happiness and pride.

Now for my turns. Here I am getting into something altogether different again. I pick out a landmark below—a peninsula on the lake marked by a small brush fire which streams thin smoke into the air, point my plane into the warm sun, and start around.

Shallow on the upwind side; steeper on the downwind. The wind is still from my right—I turn into it, bank, and the ground appears below. Wow! The point is sliding past as though someone were pulling a rug out from under me. At this rate I'll be back over Danbury by the time I've turned 180 degrees. Well, not quite—but as I come on around I can see that I have drifted quite a bit down from the lake, and the wind is still carrying me along. Steepen the bank now; over she goes. More ground comes in sight, wavering past. The steady rushing sound of air changes subtly, taking on a more urgent note; I am dropping the nose a bit. I must ease back on the wheel, bring her up, ride her around. I steal a quick look from watching ground and horizon to check my rate of climb—the needle is above the level mark: too much. Ease the wheel forward again; hold the bank; now decrease slightly as I come upwind once more. Sunlight floods through the windshield into my eyes; I've finished the turn. I level off and look below. Not bad—the point is off to my left a little way, but still—not bad. However, my altimeter reads under 2,400 feet—I dropped more than a hundred feet of altitude on that one.

Next time around I watch my needles and listen carefully for that telltale sound of increased speed that warns me I'm diving. I steepen my bank a little sooner, too, and shallow it out a little more quickly at the end. This time I hit it on the button: there is the point, directly below, and my altimeter needle shows precisely what it should.

Turning to the left, I lose it again. This time I have to start with a steep bank, since I am turning downwind. I don't get the wing down nearly soon enough, and the wind sends me kiting across country before I am halfway around. If I had been flying in a traffic pattern, I remind myself rather grimly, I would have been way out of it and the control tower would probably be hollering at me by this time. But the second time is smoother.

So I turn and bank for fifteen minutes in the wide blue sky, slowly discovering the tricks of holding my position in that vast and markless emptiness. How many times have I watched, from the ground, others doing this same thing, envying them! I never realized, those many, many times, how hard the fumbling student up above was working, wishing he had six eyes instead of two to keep horizon, ground, altimeter, rate of climb, and air speed all in view at the same time! But the exhilaration is there too, as I complete one more turn and see all these varied pieces that make up the jigsaw puzzle of my maneuver fall into place again. I can practically see my 360-degree circle in the sky.

I straighten out and fly for a little while. Danbury is behind me, but I still have it in sight; I can afford to take a little trip, pretending I am really going somewhere. Near Brewster, I spot a small pond with an island in it—a perfect spot for trying some really precise turns.

Shallow on the upwind; put my wingtip on the island. Around she comes; I pivot as though on an invisible string. Now I can see and feel the wind begin to chase me off again; I steepen the bank, and the island moves back where it should be. A little back pressure on the wheel to calm the engine's rising note, and around we come. Shallow it out as the sun comes on my face—we have pivoted a full 360 degrees. I bank her over to the other side and do the same thing that way, really enjoying myself now. It is beautiful, the airplane's firm response to my turning of the wheel; beautiful, the sense of mastery which I am acquiring over the wind.

The sun turns and twinkles off my shifting wings and I am utterly absorbed. My ties with the earth have been completely severed; I feel now like a creature of the air. Far down below me, brown fields and blue lake pivot and swing, tilting as I bank around, but I have no sense of height, no sense of distance, no relationship to anything I knew down there. I am the freest man alive, wheeling up here in the blue; I feel I can go anywhere.

I have almost forgotten where I am when I realize the time has come to turn and fly back home. The sun is perceptibly lower; I have been flying for almost an hour. Reluctantly I straighten out and look around. I've got to pick my road back to the airport.

I see it off to my right, the broad black runways looking tiny and far away. So complete is my immersion in this world of sky that it seems improbable that I am expected to land there. How do I get there in the first place? All around me is infinity, a trackless void; I could go anywhere, choose any of a million roads—which one will bring me safely home?

Three thousand feet in the air and five miles distant from the field, I make my plans. I want to cross the field at 2,000 feet precisely, then swing in a wide right turn beyond it, descending so that at 1,250 feet, the pattern height (the field itself is at an altitude of 450 feet), I will come into my downwind leg. I want to re-establish slowly my relationship with the ground; I want time to scan the sky for other traffic, and to think my way back home.

I cut the power slightly and, heading for the field, begin a slow descent. The nose dips and the whistle of the wind increases slightly. I do not move the wheel at all; this is a throttle descent, slow and easy, and I watch and calculate as carefully as though I were flying a DC-7C with eighty passengers aboard. Then, at 2,000 feet, just as I am about to reach the field, the bumps begin to hit me.

The change is quite abrupt and wholly startling. The airplane dips and dances, the air speed rises and falls. I increase power to cruising speed again and try to hold her 2,000 feet exactly as we cross the field. It is like standing on a bongo board; I have to try to anticipate each move; in short, I have to work like hell.

Far below me I make out the wind tee—it still points down the west runway. An Aeronca sails along on its downwind leg; that would be Gene, with another student. No other plane is visible in the sky. He will be landing in a couple of minutes, probably taking right off again; if I can figure this right I ought to be able to get into the pattern behind him his next time around.

Time to start letting down. I put on carburetor heat, cut the power back to 1,700 r.p.m. The engine subsides to a mutter and the whistle of the wind rises. We nose downward into a rushing, wracking, ripping river of air.

Bump, bang, bump! The hills rock crazily beneath me as I try to hold the Tri-Pacer in a shallow right bank. Up, down, drop, rise; wings rock, the air speed waggles crazily. It is impossible to anticipate all these crazy movements. If I back the wheel for a sudden drop, a rising gust hits me amidships and we soar; if I push it forward to keep the nose down, the air speed mounts toward 120 and the wind starts really whistling past. I realize finally that I cannot anticipate every bump, and concentrate on simply holding my descent and my turn as best as I can. But it is difficult; my confidence is draining away, and I begin to wish that Russ or Gene were here. . . .

Bump, bounce, bang! Damn those hills! The wind hits them and leaps upward in all sorts of wild currents, catching me and tossing me with malevolent flicks of its fingers. It begins to make me mad. Dammit, it's up to me to bring this airplane down, and by God I will! Russ or Gene can't help me now; there's only me. I see the Aeronca drifting down to the runway on final approach. If that thing can fly in this, if students can learn to fly in this, then so can I!

The hills wheel, tilt, and jump beneath me, remote and impersonal. I gradually realize that my airplane itself is equalizing its line of flight; it may drop and rise as the gusts hit it, raise a wing or lower it, but as long as I am steady on the wheel it will come back to normal. My hands are sweating, my stomach is cold, but my confidence is slowly coming back again.

I am down to 1,200 feet now and the hills are close. I have turned nearly three quarters of a circle; it is time to level off and enter the downwind leg. The road to the airport is clear now—the same familiar pattern; it is almost like turning off the four-lane, long-distance turnpike into the well-known road that leads to home.

I increase throttle to 2,000 r.p.m. and pull the nose up slightly. I want to give the Aeronca time to turn inside me and get ahead—ah, there he comes! He is probably doing no more than 70 miles per hour; I have to be careful not to catch up to him. Now he banks and turns as I swing around; I lose him for a moment; then, as I come downwind at 45 degrees to the pattern leg, I see him floating ahead of me, just as planned.

Bump, bump, bangety-bang! I feel I am losing altitude. I hold the nose up a bit more; I don't want to increase power here, I just want to fly a bit more slowly, to find the point at which I will mush through the air, neither climbing nor falling and with speed at a minimum. At 90 mph I will have a safe maneuvering speed. But it is work to hold her at that because it is an unnatural attitude for the airplane; she wants to straighten out and fly. A few turns upward on the trim tab helps, and so we mush and bounce and bang our way along the hills.

I turn on base leg as the Aeronca drifts down toward the field. Time for first flaps now; that will help. Here it is bumpy as hell, with the wind catching me crosswise. I crab her around to compensate for drift until I am mushing along really sideways. We reach the farm at the top of the hill where I turn once more on my final leg.

Cut the power now, hold the nose down. My channel of air is clear before me, and wind rushes up along it in surging waves. I feel I am too high. I want to land just beyond the intersection of the runways, but I am still way up in the air. I pull the throttle back hard, but it is already at the stop, the power cut. We are leaping around like a bucking bronco; why can't I keep that nose down? Now I see the wind sock; it does nothing to reassure me, swinging and flapping at times almost directly cross to the runway. Lower we come, and lower. I feel the cross wind now; hold that wing down! I still seem high. I am correcting constantly. Now I see the Aeronca touching down, taking off again. The runway comes up, swinging in arcs ahead of me. I still feel high. Should I apply power and go around again?

No—here comes the runway, flattening out at last. By God, I am making it. Correct for the wind; hold the wing down. Bump, gust, bump! The gray asphalt fills my vision, rises sharply, rushes past as I come down on it. Time to flare out now—back the wheel! My tail waggles as I hold rudder against the lowered wing; back now, back, back—a last quick correction; we float for an instant, then the wheels kiss down. So softly do I touch in all that wind that I scarcely notice it. One wheel first, then the other; now it is time to put the nose firmly on the ground and establish steerageway. We roll on solid land and I let out a real war whoop of joy.

Back in the office, Gene comes in as I am writing out my log with fingers that still tremble. "I held that wing down this time," I tell him. "And this time," he says, "I didn't see it. I was up ahead of you.

"Quite a bit of wind out there," he adds casually. The accolade.

On this and many other days, as I fly higher and farther, I gradually come to know the strange and wonderful ways of the ocean of air. I come to think of it as a true ocean and of myself, as I walk about my normal life on the ground, as a man who lives on the ocean's floor. I acquire, when I am not flying, the habit of looking upward, of thinking of the sky above me in terms of flight—on such and such a day, how would it be? Here is a day of dappled clouds, warm sun, and little wind—it is a thermal day, when I would feel the hot currents from the sun-warmed earth rising beneath me like invisible columns, when a river or lake three thousand feet below would be tangibly felt in the sudden drop of the cooler air above it; and when the clouds, too, would exert their living forces on me as I ride below. Here is a gray, cool day, utterly windstill, when the air would be smooth as a lake, and as delicious. Here is a day of haze and high humidity, one to watch out for—on such a day I could fly into cloud without any warning, the haze slowly thickening, the barely perceptible grayness closing around me until suddenly both earth and sky are gone.

And always, always I learn to feel the wind. Of all the air's vagaries, this for a while is the most important one to me. It governs my take-offs, my flying, and my landings; I have to learn to feel it, and feeling it, to judge it; the wind has literally to enter into my fingertips, to become as much a part of my instinct as the surge of a wave when I am swimming in the sea. And it is just as important when it is not there as when it is.

I learn this one day at Martha's Vineyard.

I am flying the Piper Super Cub on this day—a wide-winged gull of an airplane which, by comparison with the faster, heavier Tri-Pacer, seems literally to float in the air. It is a two-place, tandem plane in which I sit in the front seat behind a windshield that curves around and above me, on both sides and over my head, giving me an armchair seat on the very rim of the world. Flying the Super Cub is almost like soaring, as close to pure and effortless dream-flight as an engine-driven plane can come.

And this is a day that fliers dream about. It is sunny but cool, and almost windstill. The sock above the hangar hangs limply, barely stirring as now and then a vagrant breeze drifts by. I take off and climb into a sky of purest blue, turning in sweeping circles over the beach and the sea until I reach three thousand feet. Then, with the beach as my line of reference, I practice turns, making esses along it for a few miles, then turning to come back and make some more. It is the most beautiful feeling in the world.

Coming back at last up the beach toward the airfield, I decide to practice some landings. I give no thought to any wind there might be; there was none before. And so I fly into the same pattern as the one in which I took off, turning into my downwind leg, then base, then final, cutting my power and nosing toward the ground.

The sense of floating is ineffable. The Tri-Pacer, at this point, would be coming in at 85 miles per hour, nose pointed toward the runway, with a real sense of speed. The Super Cub drifts down nose high, some 30 miles per hour slower, floating all the way. With its three-point gear, it will land in a near stall; the flare-out is far gentler, a stage in which its forward motion, by comparison with the Tri-Pacer, seems almost to cease entirely and it just drifts along until the wheels come down.

This time, in air that is as smooth and balmy as a summer pond, it seems to float forever.

At first, the sense of exhilaration holds me spellbound. The gentlest touch on stick and rudder brings the small corrections I need as I descend languidly toward the brown sod runway stretching out ahead. The engine whispers to itself; the propeller is barely ticking over. The runway shifts slightly—left, right; it rocks a bit as I correct for the slight drift. Then I notice something odd: the runway doesn't seem to be getting any closer. It is almost as though some unseen, unheard force were trying to hold me suspended a hundred feet or so in the air.

I drift on down, as peacefully as a leaf wafting vaguely in a breath of air. The near end of the runway slides past below me; I am still a good fifty feet high. At this moment things begin to seem definitely eerie. What am I doing? What have I done wrong? Should I push forward the throttle and climb, to try again? I am extremely undecided; but there is still plenty of runway ahead and sooner or later I *must* come down.

I see myself drifting a bit toward the left and correct for it. The runway unreels past me, slowly. But now I am almost there. It is time to start my flare-out—slowly I bring the stick back, easing the nose into the air, waiting for the soft bump of a perfect landing on the springy grass.

But it doesn't come—and still it doesn't come. I cannot be more than five feet off the ground now, and still the airplane sails quietly along. I dare not pull the stick back farther; already I am almost stalled. Nor do I dare to push it forward and let the nose come down; I feel I am too close to the ground for that. I am really apprehensive now, longing for a word of guidance, just a word from someone to tell me something I never thought I would need to know: what to do with an airplane that won't land.

Suddenly, behind me, bump and thump, the tail comes down. It hits, bounces, hits again. The front part of the airplane is still off the ground and now the nose, ridiculously high, begins to swing leftward. A second stretches into an eternity while I, with a sort of paralyzed fascination, watch the long grass at the edge of the runway creep closer and begin to flow back underneath my wheel. Then the nose at last comes down.

It is a very gentle landing, crazy as it is, but by this time the left wheel is entirely off the runway. It snags in the long grass, and immediately the Super Cub starts to slew around. With the energy of desperation I try to save it, giving it hard right rudder and clamping my heel on the right brake with all the strength I have. She straightens momentarily, but the long grass is too much for her. Leftward she slides, the tail comes swinging around and we go into a wide and rapidly accelerating skid. There isn't a thing in the world I can do; I sit and let her go. She does a complete circle and comes to rest in knee-high grass, the engine still ticking matter-of-factly as if to say, "Well, Buster, here we are."

Back at the hangar, a small delegation is ready to greet me. I still have no idea what happened, and the grinning faces lined up there do nothing to relieve my total confusion. "Well," says Steve Gentle, "I see you found a new way to bring an airplane down."

I try to laugh it off. "That's just the trouble," I reply. "I couldn't *get* it down. What in the world was going on out there? She just wouldn't land. . . ."

"Take a look at the wind sock," says Steve.

I look. There it hangs at the top of the hangar, only now it isn't hanging limply anymore. It droops and sags still, to be sure, but it is unmistakably filling and alive with a gentle breeze coming from exactly the opposite direction from my landing. In the twenty minutes or so that I was flying above the beach a breeze had started up, and I had made a downwind landing.

"Well," I say, "I'll be damned."

"Yes," says Steve, "that was a real high-class groundloop. I trust you'll never land without looking at the wind sock again."

It's a thought to remember, always.

Once I am to see the entire air convulsed by an elemental riot, a howling, heaving, horrible civil war of all its parts—wind, rain, cloud, heat, cold, boiling together in that most terrible and awe-inspiring of sights a pilot in the air can see: a major thunderstorm.

It is in Texas, on a summer afternoon. We have been

flying all day across the country, Bill Strohmeier of Piper, Tim Asch, photographer, and I. For the last hour our twin-engine Apache has been slowly climbing over rising land, the flat Texas plain almost imperceptibly turning into low foothills, the clear blue summer sky clouding over with a dim and smoky haze.

We pass the town of Big Springs in waning light, keeping a wary eye out for the jets that practice in the area—dart-shaped, delta-winged creatures that climb at nearly the speed of sound, are come and gone almost before the mind can grasp their sudden presence. But we see no jets, nor any sign of life in the slowly darkening sky; we are utterly alone. And so we fly on, climbing slowly from 4,000 feet to 5,000 to 6,000, and higher still while always the ground below us rises too, like a tilted table, so that we are never more than 1,500 feet above the gray-brown, desolate terrain.

Though it is hours before nightfall, the sun begins to disappear. It fades out, loses its brilliance, then its light and then is gone. In its stead we see a flame upon the plain ahead, a long, bright tongue of it licking out of solid ground, at its tip an immense and trailing pennant of black smoke. It seems like an illusion at first, something that cannot possibly be there; it has no explanation, it is unnatural, yet there it is. As we draw near we see it is an oil well fire, spewing from the earth itself, a vision from primeval times. There are not even men to fight it; it just burns. And beyond it lies a great, fat column, a topless toadstool of rain. Lightning flashes from it intermittently, forked tongues stabbing at the ground as though to kindle other fires there; and the sky above it darkens into shadow. It is as implacable as a mountain, that rainstorm, and seemingly as solid; we give it a wide berth and fly on.

Rain spatters on our windshield now, fat drops that burst and trail off into numberless rivulets that split, and split again, flowing with eerie speed across the plexiglass. Ahead the sky is now a dark slate-blue, impenetrable, lightless. Under it, above the rising land, a haze begins to form, thickening and hardening into a cloud floor. Between the two layers we still climb, to 8,500 feet now and still higher. There are mountain peaks ahead; we are approaching Guadalupe Pass.

We climb that pass like mountaineers, thrusting our way ever upward across a tumbled field of cloud. They build and build beneath that darkening, slate-blue sky; and yet the air is deathly still. There is no sound, no motion, no hint of anything alive except ourselves; and we ourselves, in almost a dreamlike state, feel our own forward passage through this ominous world only when occasionally we brush a topmost peak of cloud and see its shreds go whipping past below.

Then, to the right, through a break in the awesome clouds, I see the 8,000-foot summit of Guadalupe Peak itself.

No Dante could have visioned such an apparition. It is a peak that seems born of the silent, massive clouds piling below and all around it, silhouetted against a brighter spot of sky. It is carved of iron, as black as the core of the world, etched and upflung against the sunless heavens. It slants darkly upward out of marbled whiteness to a sharp, many-pointed peak; then plunges, sheer, in a drop that disappears into a bottomless abyss. Behind it, as we climb, the horizon turns a pale yet livid yellow, like the gleam of distant flames.

We climb and come abreast of it; and then, abruptly, the entire landscape opens up below.

We have reached the pass, and for a moment we seem poised on the lip of the world. For the first time, now, I see the peaks below us—brown and brutal, falling steeply to a wilderness of rocks in which writhe wisps of clouds like sulphurous rivers. An immense valley stretches ahead, a desert of black and gray. Beyond it, clearly visible, the land rises as steeply again to a tremendous table top. Between that top and us, all through the valley, is a witches' garden of storms.

They are as clearly defined as pillars thrust into the desert floor. Between them the air has an unnatural clarity; we can see for miles. In them, the air boils. There is no other way to describe it—it heaves and seethes and bursts apart in flaming tongues of lightning and whipping tendrils of smoking cloud. And we, descending now, enter that labyrinthian world of clarity and storm, picking our way between.

Almost at once, the wind seizes us. We feel it bodily, a giant hand flung up to check, then speed the plane. The beat of the engines changes, rises, sinks, rises again as we soar on unseen, buffeting gusts, then drop like a stone, only to be struck another upward blow. We fight the wind and the air like a ship in a hurricane. Beside me, Bill flies with both hands on the wheel; he has lowered our landing gear to cut our speed and now he holds the airplane nose-high to slow us down still more. To our right a storm center towers, yards away it seems. For full minutes I stare into its heart, fascinated. Clouds whip up, form, are torn apart before my eyes. The black heart of it rushes and roars like solid water falling heavenward. Never in all my life have I imagined such unearthly power; it seems that the very earth could be ripped apart by such a storm.

Time stands still as we wander across that nightmare valley, flung here and there, tossed up, sucked down. The engines beat, the airplane flies like a mad thing. There is no room for terror; I feel only awe. And then we are across, climbing the ridge, soaring over the lip of the mesa and flying in the relative peace of that immense plateau beyond.

Peace, did I say? The air is still, or relatively so; we only bump and bounce now. It is the earth below which is convulsed here. We are at 11,000 feet; the land is perhaps 4,000 feet below us. And it is a running, ripping sea of water everywhere.

Countless tons of rain must have been dumped upon it by those storms. Streams run below us, mighty rivers, rippling, wave-torn lakes on which debris floats and boulders tumble end over end. We hear no sound of all the dreadful noise there must be below; we fly above it, untouched, fascinated by the unbelievable sight of a suddenly liquefied world. The sky is black above, the world is water below, and we fly on, the last living beings, it seems, upon this planet—three men in an airplane, alone.

Within the hour, we reach El Paso. The sun is shining, bright and hot, the sky a rain-washed blue. I turn for one last look at what we have left behind; haze has blotted out the storms, the mountain peaks are gone. "Five-one Poppa, cleared to land," the radio says; then we are down.

3. *Of Pilots and Pilotage*

When my good friend Bill Strohmeier was some twenty years younger—that is to say, when he was fresh out of Amherst—he went to work for the Piper Aircraft Corporation. That was in 1937. Piper at that time was producing the famous

Piper Cub, the "Model T Ford of the air," the first true mass-produced light plane. The Cub was a high-wing monoplane seating two passengers tandem-fashion, one behind the other, flown from the rear seat and powered by a 37-horsepower engine that floated it up into the sky with much the same sensation that a man would get from having himself pulled up by a balloon. A good pilot, in a Cub, could get himself a thermal current and soar like a sailplane. The Cub traveled at about 60 miles per hour, landed at about 30 (in winds of more than 30 miles per hour it could, and sometimes did, fly backward or land sitting still), and it did more than any other airplane to put wings on Americans. And Bill Strohmeier, as a Piper salesman then, a wartime instructor later, and a Piper public relations man now, has put wings on a lot of them.

He began by selling Piper Cubs. And thereby hangs a story which illustrates this chapter on cross-country flying.

Taking a trip, a *real* trip, in an airplane is one of the climactic joys of flying. It combines everything the student has learned and all the things he hopes to achieve in one truly soul-stirring package. It involves the excitement of anticipation, the satisfaction of careful planning, the sleepless joy of anxious and endless recapitulation of details, the conscious application of every acquired skill and hard-earned piece of knowledge, and the final, ultimate fulfillment of arrival at some unknown port of call heretofore identified only as a circle on a map, now suddenly become reality. To put down at last at some strange airport after a couple of hours or more of flying a planned and executed course, coping with wind and weather and one's own apprehensions and overcoming all of them—that is to have really lived.

Nowadays the resources the pilot has at his command to achieve this end are truly impressive. He has, first of all, maps put out by the U.S. Coast and Geodetic Survey, regularly updated to keep abreast of the rapidly changing airman's world. They come, for his purposes, in two sizes (jet pilots can get a third type geared to their sonic speeds)—the Sectional Charts, on a scale of 1:500,000, and the World Aeronautical Charts (WACs), scaled 1:1,000,000. The Sectionals, designed for local cross-country flying, show things on the ground in greater detail, but the WACs, used for long-distance flights, show only the more prominent features of the terrain.

The amount of detail shown on a Sectional makes the average gas station road map look like a blank sheet of paper. There is, first of all, the earth itself—all its dips and rises and mountain peaks and chains graphically presented in contours drawn in varying shades so that, with just a little practice, the student sees a mountain really as a mountain, a valley as the easily recognized trough it is. Lakes and rivers are reproduced with an exactitude that takes in every bend, every point or bay, every island and even the shape of the island. Swamps, marshes, canals, draining ditches, sandbanks and dry washes are also shown. Highways come in three sizes—small, regular, and super. Turnpikes, one of the best route indications now existing, really stand out. Hamlets, villages, towns, and cities are pinpointed with exactitude. Airports, of course, are all clearly indicated, along with their altitude, their longest runway, their radio frequencies and lights and rotating beacons, if any. The larger ones are drawn with runways as they would appear from the air; beacons with flashing codes have the codes indicated in dots and dashes. Omniranges, low-frequency radio beacons, and all other navigational aids are there with call letters, frequencies, and other pertinent information. The air's congested highways, where traffic rules exist, are outlined: air traffic control zones, control areas and restricted areas where Army, Navy, or Air Force may be practicing with antiaircraft guns, bombers, rockets, or drones. Finally, to really get down into detail, Sectionals also show radio, television, or water towers, tall chimneys, high-tension lines, isolated factory buildings, mines, monuments, or almost anything else that may stand out on the ground below and provide the pilot with a clue to where he is.

On the back of the charts, just in case a pilot is forgetful, are printed a number of pertinent rules and regulations, such as the Visual Flight Rules (visibility minimums, distances to be maintained from clouds, etc.), how-to-file-a-flight plan, ground-to-air emergency code distress signals which can be laid out on the ground with strips of cloth or tree branches or stamped out in the snow, visual emergency signals with the hands, arms, and body, the phonetic alphabet, and a complete list of all the airports shown on the chart with their locations, facilities, and particular characteristics such as unusual obstacles, landing and take-off patterns, and hours when they are attended.

On this highly detailed piece of paper (the government sells them for 25 cents each) the pilot plots his course from point to point, using as few or as many of the navigational facilities as he wishes to. If he has full radio equipment, he may fly from one omni beacon to another, or from one low-frequency range to the next, paying little or no attention to the ground. Or he may plot his course direct, using various terrain features to check his progress. Or—and this is still the best system—he may use both, following the beacons but keeping an eye out below so that in the event of radio failure or a beacon's being out of commission he will still know where he is. If he keeps a good and constant check, he can project his forward progress ahead of his plane and estimate his times of arrival at certain places, which adds considerably not only to the safety of his flight but also to its challenge and ultimate satisfaction. Any way he does it, with the equipment the modern pilot has at his disposal, a cross-country flight is not only safe but fun.

Back in the days when Bill Strohmeier started selling airplanes it was quite different.

For one thing, the Sectional Charts were scanty then. A plain Rand McNally road map was the nearest thing to them—and as anyone who has ever tried to really follow a road map knows, it left much to be desired. Nor were there any radio facilities for light airplanes. In fact, the Piper Cubs which Bill sold came with only one navigational instrument: a magnetic compass mounted on the top of the instrument panel and this was a $10 extra.

It was carefully pointed out to Bill, in those impecunious Depression days when he was given his first plane and told to take to the aerial road, that the compass was the property of the Piper Aircraft Corporation and as such was not to be included in the sale. "Be sure and take the compass out before you deliver the ship," he was told. "Otherwise we'll take it out of your hide when you get back."

Naturally in the excitement of his first sale, Bill completely forgot these instructions. Only after the sales contract was signed, sealed, and delivered did he remember about taking out the compass. "By that time," he told me, "it was too late. I just couldn't go back and tell the customer that the compass wasn't included. So I slunk home and confessed that I had sold the compass along with the plane."

It probably wasn't the first time this had happened, but Bill's luck was not running good. "That's too bad," said the plant manager . . . and they sent him out again without one. And Bill took off from Pennsylvania in his Piper Cub and for three months, back and forth across the southern states as far as Florida, he flew with one eye on a road map and the other on the ground. He flew by rivers, roads, and railways, nothing more; and when he really got lost he set down in a pasture and inquired of the nearest farmhouse where he was.

That's what is called pilotage, and it is the first form of cross-country navigation any serious student of the art of flying has to learn.

Danbury; January. I have come to fly on a windy, wintry afternoon. On the hangar roof, the anemometer hums, registering about 25 miles per hour, 30 or more in gusts, on the gauge below. Despite bright sunshine, it is bitter cold outside. "A little windy to fly the pattern," says Gene as I come in. "Today we'll try some cross-country by pilotage."

He gives me our route—Danbury to Brainard Field in

Hartford, Hartford to New Haven, and return. I sit down with the Sectional Chart and the plotter, a basic instrument of clear plastic which combines a straight-edge ruler, a compass rose with the headings marked off from 0 to 180 degrees and back again, and a mileage measurer.

With this I rule off our line of flight in heavy black pencil, then figure out the magnetic headings (plotter and chart are correlated to make this a simple matter of aligning the line of flight with a line of longitude), and finally the distance between the points we aim to fly to. I pencil the information thus gained on the edge of the chart; then I go out to join Gene, who has just finished checking out the airplane.

He takes a look at the course I have laid out—it is a good deal like a navigator's course on shipboard, except that this one travels across populated countryside rather than empty water. "Where are your check points?" he wants to know.

Check points. Aha. I hadn't thought of that. "Check points?" I say with a certain bravado. "Well . . ."

Gene takes the chart and explains.

"One thing you've got to know, always," he says, "is where you are. In order to know where you are, you've got to know how fast you're flying. You may be indicating 115-miles-per-hour air speed; at that rate you'd be in Hartford in less than half an hour. But suppose you're bucking a 30-mile-per-hour wind? What would that do to you?"

A child could see that this would cut my actual speed over the ground to about 85 miles per hour. "Aha," I say.

"The only way you can *really* figure how fast you're going," Gene says, "is to check your speed over the ground. So you mark off check points on the map. Simplest thing is to take them about ten miles apart. That makes for easy figuring. There are lots of good check points to choose from. Towns, towers, lakes, monuments, railroads, rivers, bridges, airports. See what you can find, mark them on the map, and then let's go and see how they work out."

Back to the office, back to the chart. I take a long, hard look. First of all, there is the city of Danbury—we have to go right across it. Our heading is 72 degrees magnetic; the wind, I figure, will be pretty much on our tail since it is almost due west, so our drift should not be too drastic. Next I pick out a railroad which we have to cross at Newtown. Not far away is the Fairfield County Hospital, prominently marked on the chart as an Institution. Next, a river; then another institution on our left; then the town of Waterbury which will be directly below; then New Britain (tower left, monument right); then Hartford.

I circle all these things on my map and report back to Gene. "Okay," he says, "let's go." We untie the airplane, climb in, I start her up, and we taxi out. The wind, about 25 mph and gusty, is just about straight down the east-west runway.

An Aeronca is flying the pattern; he lands as we taxi out. He comes down so slowly in that wind that he barely seems to move. He flares out late, having come in high; we pass him as he touches down—nice, a little wavering, floating to earth like a leaf. I believe I could have caught his wingtip and run alongside of him as he landed, so slowly was he moving. But he was working in there, I'll bet.

At the end of the runway I turn to face the wind, set the brakes and check out the plane. Carburetor heat off, rev up to 1,800. She is as smooth as silk, no rough spots in this engine. Scarcely a variation from one mag to another. Carb heat on; r.p.m. drops, okay. Oil pressure, okay; oil temperature, okay. Left tank full, right tank about half; we are on the right tank and stay there. I check my trim, put on first flaps, turn onto the runway, scanning the skies. The Aeronca is on his final, so I hold until he has passed over us and landed. Again he is high; he touches down far down the runway. I move out, line up, take off carb heat, and off we go.

It is a good take-off; we ride up smoothly on a strong current of air. We are climbing like a rocket and I am pushing the stick forward; I take off flaps and we climb out at about 1,000 feet per minute and 90 mph. Ahead and below I can see the Aeronca; I know where he will turn, so I wait and keep him on my left. I am to climb to about 2,500 feet, circling the airport before we take off on our course. But before I have made a full circle we are already close to 4,000 feet. Gene says this is all right, it is smoother up there and visibility is excellent. I square away on 72 degrees; the gyrocompass is practically on the nose with the magnetic one so I don't have to reset it. Almost before I know it, (tail wind!), we have left Danbury behind.

Off to our left Lake Candlewood lies like a smooth splash of silver in the brown hills. Its fingers reach into the chinks and crannies of the countryside. The sky is blue, the sun bright and white clouds ride in it, billowy, with gray undersides. Distantly, we see snow squalls—gray curtains swirling. The air is swift and smooth up here; we are warm and comfortable, trimmed out straight and level at 3,500 feet, 2,350 r.p.m., about 120 mph air speed. Gene's voice breaks in: "Now where's your first check point?"

I look ahead and down. The countryside is a jumble of brown fields, gray woods, white splotches of farms and villages, lakes and ponds of silver blue. A road crosses below; a thin sliver of a creek ahead. Railroad—railroad—how do you find a railroad in this unfamiliar, patternless vista that seems so totally unrelated to our smooth world of air? It could scarcely be more than a thread—a thread it must be, actually; a dark pencil line, perhaps with a glinting reflection of rails—and there it is. I check my map; we should be crossing it at a curve, but this I can't make out. However, the institution, the saving institution, the Fairfield County Hospital, is off to the right, approximately where it should be. We used to bring our puppies to the vet's right near there; I know the buildings.

"What's your course?" asks Gene.

I look at the compass. "Sixty degrees," I tell him. "Okay," he says, "that's it. Now we know that for the prevailing winds aloft, 60 degrees is about right. Stay on that. Look ahead now, pick out some landmark, and steer for that. Don't chase the compass; just check it from time to time."

Slowly, now, I am beginning to relax. The airplane virtually flies itself. My fingers uncurl from the wheel. I realize I don't have to keep one hand on the throttle anymore; I tighten the little knurled friction wheel that locks it in position. My legs relax, my feet rest easy on the rudder pedals. Clouds sail past, their dark undersides streaming raggedly, their shadows dappling the world below. I have picked a lake to my left, far ahead, and Middlebury on my right; on course we pass between them and Waterbury looms ahead.

There is a river running through Waterbury, and a railroad, and a stadium outside of town on the left. I look and look for that stadium. I see factories, housing settlements, the blue line of the river, the tracks expanding into their network of sidings, contracting again to a dark thread running off into the distance. I see a drive-in theater—drive-in theaters are excellent landmarks, they should be on every chart—but I don't see the stadium. I check my map repeatedly and peer over the side. Finally I realize that it is long gone; it was on the near edge of town, and I looked for it too late. It is amazing how swiftly we catch up to each of these pencil marks on the chart. Our speed over the ground must be close to 150 mph with this tail wind; yet we seem to be just floating effortlessly along.

Halfway between Waterbury and New Britain a factory is marked on the chart. I look for it, in vain. Much later I realized that the reason I never saw it was probably that we passed directly over it. And now New Britain lies ahead, a smoky splotch beneath a large and billowy cloud.

New Britain should be easy. A tower to the left, 1,290 feet high—there it is, a slender nail stuck into a brown hilltop, painted in red and white bands. Now we are up onto the town; where is the monument? What is a monument, anyhow? I have a vague picture in my mind of Grant's Tomb or an obelisk. Gene queries me: "See the monument yet?" I look and look; all I can see in New Britain is factories, houses, a river snaking through the town, bare hills, a semicircular housing development . . . "No," I say. "Where the heck is it?"

Gene takes the wheel, tips the right wing down. "You're right over it," he says. And sure enough there it is—a perfectly defined circular little park, with a decorative pool, walks crossing at right angles and a small plot in the center. Grant's Tomb or an obelisk, obviously, wouldn't have looked like anything at all from 4,000 feet up. A monument naturally is something that is easily recognizable—from now on, I won't be looking for shafts of granite anymore, but for carefully laid out little plots of ground. "Very good check points, monuments," Gene remarks, and I agree.

New Britain is a big town. Ahead of me I see a cluster of tall buildings. I keep a little to the left of them as we approach; our course lies to the left of the center of town. But New Britain seems to go on and on. I am getting vaguely puzzled. Something about those buildings seems familiar; they're pretty big, almost as big as those big insurance buildings in Hartford. I check my chart and heading again. Nothing on the chart; heading is on the nose.

Suddenly I am sure; and simultaneously Gene's voice breaks in: "Do you see the airport yet?"

"Why, that's Hartford!" I exclaim. "My God, we're already here!" I look hastily around. Just ahead of our right wing is a big airport. "There it is," I exclaim.

"Which one is that?" Gene asks. I remember that there are two airports at Hartford, one practically right beside the other. I check the chart. Aha, there is a river between them. Another look—this one is on the near side of the river. "That's us," I tell Gene, and he nods, satisfied.

Now here is a thing to remember: thinking I was still over New Britain, I might have flown right over Hartford and the airport and over the other side of town, still looking for Hartford to loom on the horizon. Speed in the air has no relationship to speed as we know it on the ground—but it *is* speed. We seem to float, droning our peaceful way along; but two and a half miles of countryside are zipping past below our wings every minute. On the chart the distance between the yellow area that marks New Britain and the yellow area that marks Hartford is about five miles—that makes two minutes. No wonder I thought New Britain was a pretty big town! If you put it together with Hartford, it is—but what I must remember is that it is as important to know when you *leave* a town as when you arrive at it. One of the easiest ways to get lost is to fly past your destination simply because you didn't realize you were already there.

Brainard Field is a big airport. We do not land here; we circle it once, and Gene points out to me a runway marked with a large white X. "That's what a closed runway looks like," he says. "Now if you see that X beside the wind tee, or at an intersection of the runways, that means the whole airport is closed." Very good; I will remember.

Course to New Haven: 213 degrees magnetic. We are heading southwest, and the bright sun is in our eyes. I have an institution marked as my first check point, to my left at a place called Rocky Hill. We will be getting a strong wind drift now; on this course that wind that helped us to Hartford will be hitting us right off the starboard bow, blowing us to the left. By my chart I see that we should be three miles or so to the right of the river, which winds along below, speckled with ice floes. I try to head away from it a bit, but I can't seem to. I keep going slightly toward it. Ahead, against the sun's white glare, I see a group of low, brown-black hills.

Well, I have a descriptive place name: Rocky Hill must be somewhere near here. I look around for my institution. Off to the left is an enormous greenhouse; that would hardly be it. I'm beginning to flounder a bit. I check my gyrocompass; it reads close to 0 degrees; the magnetic is swinging back and forth in the vicinity of 330 degrees. I reset the gyro and peer over the side again. We are over the rocky hills now, but I cannot find my institution.

"Where are we now?" Gene wants to know.

"Well," I say, "I'm looking for this institution here . . ." I point it out on the chart. "But I can't find it."

"Take a look at the river," says Gene.

"I know," I say. "I'm trying to get away from it, sort of gradually. I figure we ought to be heading away from it, slowly."

"See that big bend there?" Gene says.

I look. Sure enough, the river, in an almost right-angle curve, is sliding away to my left.

"See that town off to the right?" Gene continues. Yes, I do.

"Well," he says, let's take a look. Here, here's the right-angle bend in the river. Now here's your town—Meriden. And here you are."

Here I am indeed. Rocky Hill and the institution must be ten miles behind me at least. "Boy," I say, "you sure move along up here."

"You've got to keep remembering that," Gene says. "Now you're right on course. What are you steering?" I check: 210 degrees. "Okay," says Gene. "That seems to keep us on course and take care of wind drift. The wind may have changed, you know, since we left Danbury. Stay on 210 degrees, and it's up to you now to find New Haven."

Ahead, far off, to the left, a lake is polished silver in the distance. There is a lake on the chart too, a smallish sort of thing, right on our course. I light a cigarette, relax, and head for the silver disc that shines so brightly in the haze.

God, it's beautiful up here. Seen in this light, facing the sun, the earth below shimmers; it scarcely seems to be there at all. The plane bores ahead smoothly, straight and level, a lovely, dear thing, my friend, my Tri-Pacer for the moment, loyal, steadfast, warm, comfortable, utterly reliable. Gene is smoking, gazing pensively out of the window, lost in his thoughts. Conversationally, he says: "What are you steering for?"

"That lake up ahead," I say. I point it out on the chart. "We must be pretty near there now."

"You mean the lake with the island in it?" Gene asks.

I peer ahead against the sun. Sure enough, there is a tiny dot in the shimmery silver. H'mm.

"Here," says Gene, taking the chart. "Take a look. Now here's a lake that really makes a fine check point. See that U-shaped island in the middle of it? You can't mistake that, can you?"

I follow his pointing finger. It's a big lake, near North Branford, a good long distance away. Furthermore, it's a considerable way to the left of my course, three or four miles. I look ahead again. Yes, it *is* quite a long way off. I swing to the right again until my compass reads 210 degrees.

This is a business, this visual cross-country flying, where

you don't trust your eyes alone, or your compass alone—you trust them alternately, and you check them one against the other. Two hundred ten degrees is my heading; I can stay on that and rely on it. But suppose the wind slackens? I have no way of knowing it; if it does, I will find myself compensating for drift that isn't there. So I don't trust my compass entirely; I also check with my eyes. But my eyes must be trained to pick out the right landmarks at reasonably the right distances—in this case, I have been steering for the wrong lake, when it is as plain as a pikestaff (now) that this is a much bigger one than the one I had picked out on my chart. And with an island yet. So, I should check my compass heading; if I had, I would have seen that I was heading about 190 degrees instead of 210 degrees. . . .

Now I concentrate. "Off to the airport," says Gene, and I am determined to find it unaided. The lake with the U-shaped island is now clearly visible on the left. Beyond it is another one, an elongated silver splash. I look at my chart—a line drawn from lake to lake and beyond would split New Haven airport down the middle. Also, on the right is a large silver pool, merging into a limitless sheet of silver beyond—clearly this is the harbor. The airport lies on a point on Long Island Sound; the harbor extends right along this point. That's where it must be, there, ahead in the smoky haze.

"We're going to land at New Haven," Gene says, "so when you get near the airport, start letting down to pattern height."

We ride along for a little while. The point is now clearly visible. I check my chart again—a black line indicates a jetty extending from the point into the Sound. I look ahead: sure enough, there it is, a pencil mark on the silver.

I remember flying back one day from New Bedford to Martha's Vineyard. As we crossed the island, ten miles from the home field, I had begun my letdown, airliner style. Engine throttled back, we coasted down a long track of air. Remembering, I said: "I figure we ought to start letting down about now. Okay?"

"Can you see the airport?" he asks.

"No," I reply, "but I'm morally certain that it's right ahead, there."

"Wait till you see it," he says. "This is an airport you don't know. When you've spotted it, then you can let down and look it over."

And now I do see it. At first I am not quite sure—there is a smudge on the shimmering earth ahead, a place that somehow looks man-leveled, man-made. I fly on for a while until I am sure. Then I start to let down.

Carburetor heat on; the beat of the engine changes. Throttle back to 1,800 r.p.m.; the nose drops. Trim up a bit; now we are descending on a long, lovely, slow-moving slant along our corridor of air. Now I see the airport clearly. The altimeter sweeps slowly around—3,000 feet, 2,500, 2,000, 1,500. Overfly the field at 500 feet above pattern altitude, Gene has said; I push the throttle slightly forward and watch the rate of climb rise from DESCENT to O. The beat of the engine is a beautiful, calm, controlled sound. With the throttle alone, I can keep the Tri-Pacer flying level at about 100 miles per hour. We circle the field, and I see an airplane taking off. I check his runway against the wind tee—okay, that's it. I head out across the bay, make a wide, sweeping turn to the right, in the direction of airport traffic. "See?" says Gene. "This way, you can see the whole airport and everything that's happening on it." I can indeed—like a panoramic painting it is spread out before me as I turn into my downwind leg.

First flaps; and we slow perceptibly. Turn into final, 80 mph. The runway lies ahead, smooth, broad, inviting. A white-winged airplane is at the head of it, holding for me. Full flaps; we slow again, and I hold the nose down. We're a little high, but here comes the ground. Smoothly it rises to meet us. Wheel back for the flare-out. For a few brief instants we glide just above the blacktop, then—zup, zup—we are down.

We do not tarry in New Haven; this is a touch-and-go landing. Carb heat off, full throttle, and we are up and away again. While I circle the field in a climbing turn, the white plane takes off behind me; I follow him as he climbs, turns into pattern height, and goes on his appointed round: downwind leg, base leg, final. Me, I am off and away, at 297 degrees magnetic, headed for home.

On the way, we come close past our house. This is familiar ground, and I strain my eyes trying to locate roads, churches, ponds. All is strange. Gene points out Redding Ridge straight ahead; but I cannot line up the open fields, the white church, the intersecting highways with the country I know so well. "They're building a school here," he suggests, and I grasp at that straw. Of course we're building a school—all those town meetings, the speech I made, the time I went up there to take pictures; but where is it?

All of a sudden I see it, and everything falls into place. "Good Lord," I tell Gene, "I didn't realize it's that far along." Foundations have been laid—they are outlined, green and rectilinear, by the canvas covering the newly poured concrete. Now, just down the road, that dark clump of trees, that must be us. . . .

And there it is: the little red house, the even smaller red cottage, almost hidden in the brown woods. I dip my wing and make a steep circle around it. There's the station wagon, sitting patiently in the driveway, there—a minute brown speck—is Jenny, our Chesapeake; there is the laundry on the line (our drier has broken down, it's time we got that fixed). . . . "See it, Gene?" I say. "There it is. . . ."

We finish the circle, and I square away. I check my compass heading again, although I could find my way back by the highway now. But I want to hit Danbury on the nose. The sun has disappeared behind the clouds and it is getting on toward evening. The finger of Candlewood Lake points directly at the airport on my chart, and there it is. I circle it once, check the wind tee, and drop down to pattern height. An airplane is taking off, and I watch him go; he climbs out and heads westward into the gathering gloom.

Downwind, first flaps, base leg, final. I know it by heart. Here, where I turn into my final approach, a barn was burning last week, a vivid red ember. Full flaps; we come in over the highway. The wind is dying; the air is soft and smooth. I ease back on the wheel and gaze ahead at the runway. Now we are almost down; wheel back; we lift, float, touch, and are on the ground. Tri-Pacer 8966-Delta coasts down the runway; the flight is over.

This is the process of learning to fly: to repeat and to repeat, to fly and fly again. Every flight is different, with a new challenge, new interest, new conditions; and yet basically the problems are the same. Thus it is a process of acquiring instinct, and refining instinct with skill and knowledge. How important the instinctual part of it is needs constantly to be re-emphasized—for in the euphoria of flight the student all too easily forgets how far up in the air he is, how abruptly his entire situation may change by the whim of wind or weather or, the ultimate emergency, an engine failure. Nine times out of ten the engine failure will be human in origin—a gas tank run dry because fuel consumption has been forgotten, dirt in a fuel line that was not checked and drained in time, a wire loose that would have been detected by a conscientious eye. And nine times out of ten, too, the abortion of a flight because of weather—a frightened turning back, or a forced landing in the face of minimum conditions—will also be the pilot's fault, because he failed to see in time what he was getting into, and got lost or wandered helplessly in fog or cloud until he had to come down.

One way or another, the ground will always claim the airplane again. This fact must always be remembered, and provided for. Somewhere a flight must end—it is the pilot's duty to see to it that it ends at a time and place which he has planned for and in conditions in which he can bring his aircraft safely down.

I think of this particularly in relation to my next cross-country flight. This one very quickly resolves itself into a battle with the wind. It is a flight in which, more than once, I find myself drifting helplessly across the face of the earth, uncertain as to where I am.

The flight begins with a long, slow, climbing turn to 3,000 feet above the airport. We leave the north runway against a gusty head wind; now on a clear and icy winter's day we are bound to Great Barrington in Massachusetts, a grass runway airfield about fifty-five miles due north from Danbury. From there we will fly to the Dutchess County airport on the Hudson near Poughkeepsie, and then home.

Mindful of my recent flight with Gene, I have marked off my course in ten-mile segments. Russ, who is flying with me this time, gives me a simple formula for computing my speed above the ground: time my progress over a ten-mile stretch, then divide the time into 600. Since we have a strong head wind, my first problem is to find out just how strong it is, and to figure then what my true ground speed is.

At precisely 2:49:30 by the cabin clock I cross over the airport at 3,000 feet and head north on a magnetic course of 13 degrees. My first check point is a peninsula jutting downward into Candlewood Lake. Wind or no wind, the air feels as smooth as glass. There is no drift; we are flying arrow-straight. The earth and the sky are beautiful: low black and brown hills looming ahead, strips and layers of distant clouds at various heights, and off to the left, like a barrier against the horizon, the dim, dark shapes of the Catskills.

At 2:56:30 exactly I cross my peninsula. Seven minutes for ten miles. I figure in my head while Russ checks with the computer: approximately 86 miles per hour. We are indicating about 127, which means a wind of about 40 miles per hour from just about dead ahead. We have 45 miles still to go. Here is where I learn about ETA, that all-important Estimated Time of Arrival. Russ figures it on the computer—we should arrive over Great Barrington airport at 3:27.

Meanwhile, I learn more about map reading.

"Where are we?" Russ wants to know. I look over the side. I know this road; it's the road along the Housatonic River, past Cornwall to Lime Rock. Road and river are bending now to cross our course from right to left. I put my finger on the map. "Here," I say. Russ wants to know why I think we are here. I point out the bend in the river below. He points out, additionally, the bridge at Gaylordsville, just off to our left. Bridges are very good check points, he says; they don't get covered over with foliage in the summer, they don't get covered with snow or ice in the winter. "You can always see a bridge."

This brings up something to remember: a good check point in summer may be a bad check point in winter. Lakes, for instance. Russ points out how easily a frozen lake, covered with snow, could be mistaken for a snowy field. Roads, on the other hand, which might be hidden for long stretches by green trees in the summer, are clearly visible in the winter, even more so if there is snow on the ground.

Ten miles farther along, he asks for a fix again. "Tell me where we are, and give me three reasons why you think so," he says.

I can give him one reason: we are crossing a highway between Cornwall Bridge and Sharon. But for the life of me, I can't give him more. I can't see the village of Cornwall Bridge; it is hidden beyond a ridge on my right. I think perhaps I see the village of Amenia, but I'm not sure. I ride along, looking. I don't know whether to look at the map first and then check the ground, or vice versa. He points out two more reasons, finally—a small lake with a creek leading out of it on my left, a valley coming up ahead. A couple of minutes later I see a landmark I know by heart—Lime Rock sports car race course, below on my right. It looks tiny, a threadlike ribbon in an irregular oval, with the Bailey bridge across to the infield—and there is the straightaway, improbably small, yet a place, I know, where a Tri-Pacer can land. . . .

We are leaving one chart now and coming onto another. We are flying in a beautiful half-light; the sun has disappeared behind a cloud layer to the west; long, distant strips of cloud lie clear and cold ahead. Parallel to our course rises a high, brown ridge that reaches almost to our level (actually, as the map shows, it is 2,600 feet high, which is one of the reasons we are at 3,000—"always find your highest point along your course," Russ said, "and plan to be *at least* 500 feet higher, preferably 1,000 feet higher"). At its highest point there is a fire tower; and I see it now, a pinhead on a hog's back, incredibly remote and isolated. It is an extraordinary feeling, flying like this alongside of a mountain. I have a valley feeling, because the valley lies below and we are flying, in a sense, in it; yet I am higher than both valley and mountain, can look down on both, and yet feel the safety, the familiarity of the one while strongly feeling the lonely loom of the other. Now the ridge bends away, the valley widens, and here somewhere, I know, must be the field.

Grass airports are not easy to find. I look, check the map, look again. It must be dead ahead. Still, I can't see it. I look for a hangar and parked airplanes. I check the map again; we *must* be almost there. Yet I haven't found it; one possibility I have been watching turns out to be just a pasture. Then suddenly I spot it; we are almost over it.

Russ says: "Supposing you hadn't found it?"

"Well," I say, "I guess I would have circled and looked. I knew we must be just about there."

"Look at your clock," says Russ. I did—it read 3:27 on the nose. "For God's sake," I said. "I forgot all about our ETA."

"That's why an ETA is useful," Russ points out. "If you can't find the field, you look at your time and when you hit your ETA, then you circle and start looking. Now a field like this is hard to spot. But look at the map here. There are a whole lot of things to go by. Over here, to the left, is the town. That's where I would start looking. Circle it—now here's a road leading out of it. See? That road goes right past the airport. So just fly along the road. Also, there's a river that goes right past it. And two lakes to the north of it, one on each side."

We turn over the airport and I take up my new heading—228 degrees for Dutchess County field. Here the battle with the wind really begins. It is coming from behind and to my right now; it is pushing me forward and sideways. I have to find out how hard it is pushing me off course, and how much to compensate for it. I ought to point higher—but how much? "Try ten degrees," says Russ. I try that heading, but I have to reset my gyrocompass, and my magnetic one is still swinging wildly from the turn. By the time it comes to rest and I have reset my gyro and taken up a heading of 238 degrees, we have covered quite a bit of ground. I check my map, then look over the side. We ought to be crossing that high ridge at a long angle, but it's very hard to see from here just what that angle should be. I check the map again. . . .

"Where are you?" Russ wants to know. I shake my head helplessly. He points to the map. "Here's a tongue of ridge sticking up, pointing north. There's a valley between it and the ridge we're supposed to be flying over. There's a town right in the middle of it—Boston Corners. Our course goes right over it."

Now I begin to see how this thing works. Look at your map, find a place, *then* look for that place. I have been going back and forth between map and ground, getting thoroughly confused, really studying neither. Still, I can't find Boston Corners either. But now I do see that tongue of high ground; I follow it northward and there is a village, away off in the distance to my right and almost behind me. . . . I must have drifted five or six miles. So I have my fix, and Russ tells me to turn and head for it; when I am there, I square away again on my course—and right away I am lost again.

This is confusing country. There are few good landmarks, actually; all I see is a confusion of hills and ridges, a haphazard sprinkling of lakes and ponds. I think I am somewhere near Pine Plains, but quickly learn that I am way beyond it; this wind, in addition to pushing me sideways, is also pushing me ahead at a hell of a clip. I go back to my futile routine of looking at the map and looking at the ground with a mounting sense of frustration. Finally I give up. "I don't know where I am," I say.

"Well," says Russ, "let's really look at the map. You know

you're somewhere around here"—he points—"so you want to find something around here that you can go by. See anything?"

"There's a town," I say. "Millbrook."

"Right," says Russ. "Not only a town, but a town with a lake to the north of it and a river leading into and out of the lake, a river with two big bends in it. Also a good-sized crossroads just south of the town—two big roads making a four-corner. That's what you want to look for. So let's start looking."

I look. I can't see anything like that anywhere. It ought to be off to my left, I should think, but off to my left is nothing but hills and emptiness. "So let's start circling," says Russ, and he banks the plane steeply to the right. "See anything?"

There it is, right below me: town, lake, river, crossroads, and all. Once again I must be at least six miles off course. "You'd better point up still more," says Russ. "With this wind, we're really drifting."

Now he points out a big highway ahead, the New York State Thruway. Also, two small lakes just before we should reach it. Squared away again, I head for them; they duly appear, then the Thruway, and now I am flying into bright yellow sun and can see the Hudson shimmering ahead.

I know I'm on course now, but Dutchess County airport is going to be hard to find, flying into the sun like this. But I'm learning. I see by the map that two bridges cross the Hudson at Poughkeepsie; they should be off to my right pretty soon—and there, silhouetted against the silver water, they are. I'm getting close. I see what looks like a big airport off to my right, but it isn't where Dutchess ought to be.

Is there another field, a military field, perhaps? I check the map—no. Now I see this is a big complex of factory buildings. We *must* be getting close. Russ's voice breaks in: "There's one real blind spot, you know, on a Tri-Pacer," he says. . . .

I think for a minute. "Yes," I say, suddenly remembering Millbrook, "Straight underneath you." "So," says Russ, "we bank the plane hard and look——" and there below is a great big airfield, runways and taxiways, hard-surfaced, hangars, planes. . . .

The last leg goes better. Russ points out a few really good check points: Hopewell Junction, with a junction not only of highways but railroads; a prison off to the left, Stormville airport right next to it with its brown dirt runways. We pass them all, and I get a really good fix for my compass heading; about 332 degrees. He points out others as we approach a whole confusion of lakes—a double railroad which crosses our course, with a pond between the two tracks (we go right over it, and again I have to tip the plane to see it, but there it is). I am also learning that landmarks which you overfly directly can often not be seen until you're right over them, so it's a good idea every now and then to heel way over and have a look. We leave Brewster off on our right, and now I know we must be getting almost home. I look for Candlewood Lake, which by the map should be pointing straight down at the airport at Danbury. There it is; I follow its pointing fingers, and there is the airport too.

We are home again, just at sunset; it is twilight on the airfield as our wheels touch down. We have been flying an hour and a half, but to me it was an epic journey. I feel like a sailor returning safely to port after an arduous voyage, filled with the same emotions, mingled of respect and awe and satisfaction of having fought my battle with the sky, and won. I was lost, and found myself; I was drifting, and I made my course; now I am home again.

The truth is that the relationship with mother earth is always very real. It may not seem so, as one bores along, 3,000 or 5,000 feet in the air, with the earth so remote, so tidy, so beautiful and touching with all its rough edges smoothed, its ugliness diminished, its problems miniaturized from the high aspect that a pilot habitually enjoys. The flier drifts across its face, and very easily can fall to dreaming—it doesn't really matter, at this distance, where he is. But of course, it always does.

I think of my friend Bill Strohmeier, without even a compass, flying alone in a Piper Cub across those miles and miles of air and earth. I have a high respect for that kind of flying. Any flier should have, for this is the kind of flying any one of us may at any time be reduced to, and it is well to know it. Not only for reasons of safety but because of an eternal truth: the flier is never so far away from earth that he can give himself up to his new element and abandon himself to dreams. Sooner or later, he must come down.

4. *Flight by Omni*

Sights and smells and sounds combine to make up memories, the lasting pictures of the mind which, at a glimpse, a whiff, a whisper, can be conjured up as sharp and clear as in the instant when they happened. A sight that lives forever with me is a cloud-hung canyon of sky over the island of Cyprus during the war, a well of clarity through which we could at last, after hours of misty wandering, descend in safety to the earth below. A smell I will never forget is the first smell of the new Tri-Pacer in which I eventually found my way into the sky—One-three-Delta, with her mingled odors of new plastic, butyrate dope, carpets, rubber, oil, and gasoline, like the exciting smell of a brand-new car bought off the show-room floor, only infinitely more so. A sound that I can conjure up at almost any moment is the music of the omni—the rhythm of the radio beacon signals which, like a shaft of light in darkness, signal to me in the empty heavens that I am steadily on course and flying home.

Omni is the greatest thing in aviation since the first cross-country pilot in his pusher biplane took a last, long look at whatever maps he had, memorized his compass headings, swallowed hard and took off, hoping that wind, weather, and his own acuity would eventually combine to bring him somewhere near where he wanted to go. It is a radio navigation system tailored to the needs of airline giants, yet available to and workable by any student pilot flying the smallest, omni-equipped Cub. It is a system so simple that it can be learned, actually learned and used, by anyone in a half-hour's practical experience; yet so complete that it can tell a pilot flying blind and with every weather factor conspiring against him exactly where he is and how to get there; and, within an instant, *when* he gets there—for he sees a needle swing and he knows that he is over Point So-and-so as surely as though he saw a signpost in the sky.

Omni—the name is abbreviated from omnidirectional

range—is rather like latitude and longitude made visible; and, as a navigational aid, its principle is much the same. Latitude and longitude are imaginary but mathematically accurate lines drawn across the face of land and sea, each with a constant reference point—the poles for longitude, the equator for latitude. By drawing in the sun and the stars, with their fixed relationship of movement to the earth, the navigator can calculate, and project onto a map, precisely where he is at any given moment. Omni works the same way, except that its lines are far from imaginary; they are man-drawn by radio across the land, and can be instantaneously made audible and visible. Any airplane equipped with an omni set can pick them up anywhere and follow them to their point of origin, a very high-frequency radio range.

The word, and the system, have been a mystery to me for a long time when I first take off and encounter omni in practical experience. I have seen the set often enough, on short charter flights with Steve Gentle from Martha's Vineyard to Nantucket, New Bedford, or Providence, ferrying passengers to railroad trains or other earthbound errands. But I have only half-understood it. It is a system that sends out 360 radiating radio signals from a central transmitting range—one radial for each point of the compass, the whole network spider-webbing out across the sky. The pilot approaching Nantucket from Martha's Vineyard, for example, over the empty sea, will be flying a course of 130 degrees magnetic. He tunes his set to the Nantucket omnirange at 112.7 and the signal comes beeping musically through the speaker: .— —–. —.—, spelling the letters ACK Nantucket's radio identification. He then rotates a selector dial marked off, like the compass, in 360 segments, turning it to 130, the course that he is flying. At or about that point a needle on a blue and yellow segmented dial becomes active and starts swinging. When it is centered, hanging quietly straight up and down where the blue and yellow dial sections meet, the airplane is on course straight for the omnirange. All the pilot has to do is hold the needle there. If it swings to the right, he changes course to the right until it is centered again; if to the left, he turns leftward, correcting when the needle shows him he is on his heading once more. And when he is directly over the range, the omni set will signal that: the needle then swings all the way to left or right, fluctuating wildly for a few moments until it has picked up an outbound radial, and a small sign in a window in the dial changes from TO to FROM.

This much I have in mind as we climb out one autumn cloudy morning from Teterboro airport in New Jersey for my first real omni flight. Larry Hoppe is with me, an ebullient young man who flies with an inborn instinct—an instinct he sometimes loudly wishes to God his students would acquire someday, soon.

It's a bit too hazy for comfortable pattern flying, so Larry has decided to give me some rudimentary omni training instead. We plan a triangular flight, from Teterboro to Caldwell, thence to Colt's Neck, then back to Teterboro again.

First off, we tune the set to the proper station. The beeping signal sounds and is indentified. I have an approximate course—it does not take into account the winds aloft—and I turn the selector dial to it. The needle centers, and we are off and away.

The rest would appear simple. All I have to do is keep that needle centered. "Steer into the needle," Larry tells me. "If the needle swings right, turn right. If it swings left, turn left. Don't turn sharply; just take a bite at it—say ten degrees—and hold that until the needle is in the center again." I am flying a 300-degree heading, slightly north of west. I try to hold it, and almost immediately become confused. If I hold 300 degrees by the compass, the needle goes off center. If I recenter the needle, my compass course goes off. We fly on through the hazy sky, and within five minutes I don't know what I'm doing.

"Center the needle!" Larry says. It is away off to one side; but so is my compass, which reads somewhere around 320 degrees. I swing back toward the needle, and the compass goes off still farther. I don't understand this at all. I'm supposed to be riding a beam in the sky, but I can't get the needle and the compass to jibe. If I get one right, the other goes off; it seems impossible to get the two together.

Larry tries to explain. "Look," he says, "it's like a railroad track. You ride along it; if you start going off it, you get back on it again. You've picked your track; just stay on it."

"But look at the compass!" I reply. "I'm supposed to be flying along a 300-degree radial—it says so right there on the selector dial—and the compass reads 320 degrees. I don't get this; am I supposed to turn the dial to 320 degrees, or what?"

"You stay on the 300-degree radial," Larry says. "Never mind the compass. You probably won't be able to stay right on it all the time; what you do is to make shallow S-turns back and forth across it—after all, you've got wind drift up here—but you stay on it."

"But what about the compass?" I ask again. "Why doesn't the compass read 300 too?"

The light of comprehension dawns on Larry's face. "For Pete's sake!" he explodes. "Didn't anybody ever tell you about wind?"

Wind! Wind on my wings, the flier's enemy, the flier's friend! How could I forget the wind! How could I forget that an airplane almost never travels in a straight line; that almost always it is crabbing, one way or another, to make up for wind! We have a wind, of course, up here; it is blowing from the right, trying to drift the airplane off to the left. I have to turn into it to hold my course. My nose, and hence my compass, is pointing up into this wind; naturally I am not flying 300 degrees straight and true, but traveling on a 300-degree radial slightly sideways. Wind correction—I had forgotten it entirely, and here I was making it automatically by pointing the airplane in the direction that would hold the needle in the center.

"I'll be damned," I say. "You mean this business automatically corrects for drift?"

"Of course it does," says Larry with the sarcasm of the expert for the innocent. "It can't help but do it. All you have to do is keep that needle centered on the dial. You can forget the wind; to hell with the wind, just keep that needle centered."

We fly on for a while in silence while I digest this revelation. It is truly startling to me. If I fly by omni, I automatically and almost immediately get my wind correction angle, that elusive bit of information that otherwise has to be laboriously figured out by plotting a wind triangle—one line for the magnetic course to be flown, another line for the direction of the wind, both lines marked off in knots or miles per hour, then connected by a third line that gives the angle by which the course must be corrected to fly true. Up here, all this is eliminated. I know my course, and set it up on the omni. I hold the needle centered, and observe what my compass reads. The difference between the two is the angle by which I must correct for wind, in this case the difference between 300 degrees and 320 degrees—20 degrees.

Given this, I could even figure out—if I had to—how strong the wind is and precisely from what direction it is blowing. As it is, I don't; all I have to do is keep the needle centered. But if the omni blew a fuse or a tube or something, I would have the information and could still hold a straight course on the compass alone.

The flying, however, is getting trickier and trickier. As we sail on, the needle tends to wander more and more. I correct and recorrect, trying to hold it in the center. "What's going on?" I finally ask Larry. "I can't seem to hold it there at all anymore."

"We're getting close to the station," he replies. "The closer you get, the narrower the beam is. You're not just flying a railroad track anymore; you're flying a track where the rails get closer and closer together. Watch now," he adds. "Watch the needle, and then look over the side."

The needle seems beyond all reason now. It is away off to one side. Suddenly the word TO in the little window flickers, then flips up and disappears. A red flag shows momentarily; then the word FROM wavers into view. Simultaneously, Larry takes the controls and banks the airplane far over to one side. "There," he says, pointing out the window at the ground below.

And there it is—a little round structure in an open field, painted in red and white checks with a pointed Chinese hat on it. "There's your omni," Larry says.

On the return leg I try to fix the picture firmly in my mind: know your course to the omni station, set it up, center the needle and hold it there. In theory I think I have it, but the last link of practical comprehension is still missing. I still don't quite understand that picture of the railroad in the sky; it fits the theory but not the practice. The final understanding awaits another day.

That comes, a few weeks later, on a flight with Bill Strohmeier down to the Piper plant in Lock Haven, Pa.

We are flying in Bill's Comanche, which to a Tri-Pacer student like me is rather like graduating from a Ford or Chevrolet to a Mercedes or a Jaguar. A four-place, 250-horsepower airplane of advanced design, the Comanche is the latest thing off the Piper drawing boards, a beautiful, fast ship with the lines of a fighter plane. We take off and start climbing; at 2,000 feet Bill turns over the controls to me. "It's all yours," he says. "Let's see how you get us there."

This is an omni flight, from range to range across New Jersey and Pennsylvania to Williamsport, Pa., with pilotage only the last few miles from there down the river to Lock Haven. We have laid out the course beforehand: first to the Hugenot omni, then to Wilkes-Barre, then down a short dogleg to an intersection called Avoca, then to Williamsport. We are using a different kind of chart for this, one in which there are no indications of what is on the ground at all—no rivers, cities, towns, or contour maps, just radio ranges, beacons, and airway routes drawn in between.

In a way, this chart is easier to fly by, as long as I am flying on omni alone. It gives a vivid mental picture of all these lines that have been flung across the sky, a sky that is not empty and voiceless as it was when the first airplanes flew, but is crowded now with aerial highways running arrow-straight from point to point, intersecting each other, branching off, diverging, converging, each with a different signal, each thus clearly defined. We see none of this, of course, as we climb to altitude in the quiet, silvery air under a high overcast; we do not even see another plane. But we are on our highway, a straight blue line on the chart from Westchester County airport to the Hugenot omni fifty-eight miles away, and it is up to me to stay there.

Bill gives me the railroad track principle again. It still does not quite fit the picture I am trying to form in my own mind. When I think of a railroad track I think of wheels that run along it and that can't get off without wrecking the train. I explain this, and my difficulty in understanding how the railroad theory fits the fact of drifting constantly away from the track you are supposed to be on.

Bill modifies the picture for me. "It isn't as though you were *on* the track," he says. "It's more as though you were flying along over it. You can visualize it down below there, and you are following it. Sure, you'll drift off it now and then, so you steer back until you're on it again. When you do drift off it, you don't turn sharply to get back over it again; that way you'll overshoot it and be off it on the other side. You just angle back toward it until you've hit it; then you correct to stay on it. It's as though you were on a spur switching back onto the main track."

That way the picture makes a little more sense. I settle back and concentrate on trying to hold over my railroad track.

The Comanche climbs smoothly and powerfully, like a car mounting a long grade. The air speed reads 120 miles per hour, and yet we are going upward at 1,000 feet per minute. Below, the Hudson River slips by, and a countryside speckled with houses gradually gives way to rolling hills of rusty winter trees and forest green. But I scarcely look at the earth; I keep my eyes fixed on instruments and gauges. I don't know yet what the wind may be doing to us; I try to check my drift by comparing the omni needle and the gyrocompass. With the needle centered, the compass seems to hold steady on 305 degrees; okay, I'll fly by that.

It is next to impossible, of course, to fly by the needle alone; it isn't sensitive enough for that. But I understand now how needle and compass go together, and so I fly in normal fashion by the compass, checking the needle now and then. Thus, after a while I find that I am drifting off to the right a bit. I start taking a bite at my compass heading—20 degrees, let's say, which makes it 285 degrees. I make a slight turn to the left; the Comanche swings obediently around; at 285 I steady her, straighten out, and hold that course.

The needle, for the moment, stays where it is. I am tempted to turn some more, but Bill says no. "You'll only cross your radial at an even sharper angle if you do," he warns me. "Remember, you're angling back in toward that track now. You're not going to get on it right away. Give yourself a little time."

Sure enough, now she starts to swing. Slowly the needle creeps back toward the center. Very slowly, as it creeps, I start correcting my heading back toward 305 again. I would like to be precise about this, and be back on my correct compass heading just as we get back on the beam. Take ten degrees, let's say, and hold that; now here she comes.

A good heading, it develops, is about 295 degrees. That seems to compensate adequately for wind drift. I don't succeed in holding the needle always centered; we are still angling gently back and forth across our line, but we are generally holding our course. My corrections grow tighter and more frequent as we approach Hugenot (the track is narrowing below)—and here it is now: the needle swings, the little red flag comes up in the window, TO changes to FROM, the range is below.

At Hugenot I must turn to take up a new heading, 287 degrees to Wilkes-Barre. It is like coming to a crossroads in the sky where I must turn left and take a different road. I turn by the compass to 287 degrees, change the course selector dial on the omni set accordingly. I am now outbound; the sign in the window reads FROM. Slowly the needle centers as the set picks up the outbound radial, and soon we are squared away again.

I can ride this radial for a long way, fifty miles or more, but sooner or later I want to pick up Wilkes-Barre and ride its radial in. The question suddenly arises in my mind: Which radial do I take?

Here again I try to form a mental picture of the system;

and again I am confused. On the chart the omniranges appear like so many little compass roses, each pointing to magnetic north. I am outbound on the Hugenot compass rose on 287 degrees. On the Wilkes-Barre compass rose I am inbound on the opposite leg, on what would seem to be 107 degrees. Should I, then, turn the selector dial to 107?

By earthbound standards this would seem to be right—you head away from one town, in toward another. But Bill quickly points out the flaw in my thinking. I am heading westward; the 107-degree radial points eastward, in the opposite direction. If I tune that one in, I will get a backward reading; the needle will be centered, but the indicator will show FROM while I am actually flying TO. My heading, however, is and remains a westward one. So I tune in Wilkes-Barre, leave the course selector where it is on 287 degrees, and sure enough, the needle centers and the little window shows me where I'm going: TO.

In brief, omni radials are *all* outbound. They do not point a plane in; they throw *out* a net along which the pilot can *guide* a plane in. In plotting a course from one omnirange to another, it is always the *outbound* heading that counts—hence the pilot draws his line of flight through the center of the omni's compass rose and takes the heading beyond it which points in the direction in which he is flying.

Down below the landscape flows past; I don't see it at all. I am now completely fascinated by this instrument flight. I watch the air speed for level flight (hold her at 170 mph), the gyrocompass for direction, the omni needle for track. If the wind is trying to blow me off course, I do not care or even know it; I just center the needle, then check the compass to find my correct compass heading. In this case it turns out, after a few gentle S-turns back and forth across the track, to be about 292 degrees. On that heading I can keep the needle almost motionless, centered on the dial. Now and again it may go off a bit, but a slow correction always brings it back again.

And now a picture slowly forms in my mind, at last, of how this works. I am high in the sky, floating across this limitless ocean of air. There are no landmarks here at all, only the clouds and, far above, the sun which, veiled behind its curtain, is more felt than seen except when it occasionally breaks out through a rift and blazes like a brief fireball in the western sky. It is like drifting aimlessly across a quiet sea—and for one who did not know the needles and the dials, this is what it would be.

But our drift is not aimless. Long ago man sensed a great, invisible presence to the northward which had the extraordinary quality of pulling a magnetized needle toward it—the North Magnetic Pole. For centuries men have guided ships and airplanes by that vast and formless and compelling mass, never seeing it, reckoning with it only in the abstract, scarcely even wondering what it was. Up here now, in this little airplane cabin, suddenly the North Magnetic Pole becomes a real concept to me.

Everything I fly by is oriented to it, beginning with the little magnetic compass sitting up there on the panel. By that I set the gyrocompass, resetting it every twenty minutes or so because, unlike the magnetic compass, the gyro has a tendency to wander. The omni beacons, too, are aligned to the same magnetic forces. All take their direction from it. When we are heading straight toward that invisible presence, all our instruments will be on zero—magnetic north. As we turn away from it, so our needles turn.

It is so real to me that I can almost see it, like a great mountain rearing up into the northern sky.

And now I fit the omni into the picture too. Below the great magnetic mountain, but carefully aligned with the invisible lines of force which it sends out across the entire world, we have drawn a spider web of intersecting radio lines. Each one of these has a central hub from which, like the spokes of a 360-spoke wheel, it radiates outward. More than six hundred of these wheels have now been set up across the United States, so that seldom is the pilot out of reach of the spider web. All he has to do, figuratively speaking, is to reach out over the side of his airplane, snag a strand and, letting the strand slip gently through his fingers as he flies along, follow it to the center of the wheel. He can find out which strand he has by simply listening to the signal and tuning his course selector until the needle centers. No matter where he is, there is always an omni somewhere that can guide him to some known, identifiable spot.

That is my analogy. My omni antenna is my finger; it picks up the strand and, watching the needle on the dial, I can practically feel it in my hands. If I go too far off course, I will lose it; I mustn't stretch it too far.

A little later, I have a clear-cut visualization of this. We have passed Wilkes-Barre and are on our course to Williamsport. Great brown hills are below us, cut with deep gorges. They rise to half our height, and the air grows rough and potholed. We have a strong wind drift here, and I am hard put to it sometimes to hold my heading. I swing back and forth across my track, trying to hold on to that slender strand of radio beam made visible on the omni dial. The compass swings; I correct and recorrect the airplane to hold it at 277 degrees. And now we come to the last great rampart of the hills; ahead I can already see the plain beyond them, peaceful and golden in the rays of the slowly sinking sun. The needle is right on the nose at this point—and then suddenly, in a clearing on that last great hill below me, I see the omnirange.

This time it is truly like a revelation. We pass directly over it and the needle swings away up to the left; swings, flutters, then centers again. We know where we are now with absolute precision; we have followed our strand to its source; we are as good as home.

5. *The Big Time*

Spring days, May days, vacation days—flying days. The world is transformed after the hard weeks of winter. The air is soft, caressing, warm. Baby buds fluff the iron branches of the trees with a soft blush of green, and the wind, the wind that sometimes frightened me on winter week ends with its hard, cold breath, now flows in gentle breezes down the valleys and over the hills. Here, in the merry month of May, begin three days of a delirium of flying, three unforgettable days in which I try my wings, full-fledged, in the airman's great, wide, challenging world.

TUESDAY, MAY 5.

My course is to Windham Field in Willimantic, Connecticut, by way of Hartford; thence to Westerly, Rhode Island, on the shore of Long Island Sound; from there a long leg back along the coast to New Haven and home.

This is the day I learn a lot. It is my first cross-country flight alone, over two hundred miles of flying in which everything is up to me.

I plot my course and calculate my headings carefully. I scribble notes on the border of the map and add various reminders—reset the gyrocompass at regular intervals, switch tanks to the fullest tank before landing, check field altitude at the airport of destination, and so on. It all looks good. Later it will look less good; in fact, this will be the last flight I will make for which I prepare such relatively hasty and haphazard notes—but I don't know that yet. . . .

I am flying Danbury's blue Tri-Pacer, N-7226-D. This is not the regular cross-country ship—66-Delta was flipped on its back and damaged in a baby tornado ten days ago and is now being repaired. My blue bird has only a basic instrument panel—air speed, tachometer, altimeter, fuel, temperature, and oil pressure gauges, a magnetic compass. There is no gyrocompass, no turn-and-bank, no rate of climb indicator, no artificial horizon. I do have a Narco Superhomer with omni, four-channel transmitter and receiver. This, too, is the basic set—no fancy frills, a far cry from the Omnigator Mark II in the Comanche, with its 27 channels, ILS localizer for instrument landings, and all the rest.

I check out the airplane carefully. My wife, Behri, has come down to see me go; her presence adds to my feeling of departure on a real, big-time journey. She waves me off, I taxi out, check the mags and carburetor heat, and fly.

I feel thrilled, elated, and apprehensive all at once. This trip is farther than I have ever been, even in a dual cross-country. I have flown the first leg before, once in each direction to Hartford, but have never been beyond there. And I will be making two landings at strange airports; I will have to get my logbook signed, take on gas (probably at Westerly), and find my way around all alone. This is the big deal.

Tri-Pacer 26-Delta is a sweet machine. She is lively and powerful and smooth, and we climb off the west-east runway like a rocket. I make my left turn, then head out of the pattern in a shallow, climbing turn to the right, get my altitude of 2,500 feet as I recross the airport, and head off into the blue.

Less than five minutes later I am lost.

I started out merrily, following my well-marked map, picking up my first check points at Newtown and Sandy Hook (the hospital, the new bridge across the river), and then suddenly I realized that I was flying the wrong leg. This was one I had laid out for Saturday's trip on which I would have been returning directly from Windham; now it has already put me several miles south of my intended course. I had asked myself already once before whether it would not be better to erase old courses from the map; now I know. *Always* erase them! Flying, concentrating, thinking of a whole lot of things at once, it is too easy to get confused and take the wrong leg. There is enough to think about in the air without remembering to make distinctions between different trips marked on the map.

I discover my mistake as I cross the Housatonic River. Without thinking this thing over at all, I decide simply to turn north along the river for a way until I have reached my correct track, then turn right again on my heading and fly as intended.

I make the turn and proceed upriver. I am trying to find check points on the map, but there are none. The distance is too short; there is only the river, and I am flying along that anyway. So after a few minutes, by guess and by God, I turn right, heading northeastward again, and hope that I have judged correctly.

Ah, wayward, thoughtless hope! The countryside below is a lovely, peaceful panorama of nothing at all. The sun is bright and warm, the air is just a bit bumpy, enough to keep me working, and there is the slightest veil of haze over all the land. I try to hold my heading, but the swinging of the magnetic compass makes this difficult. If I watch the compass too long, I drop a wing. When I straighten my wing, the compass is off again, owing to the slight turn of the airplane. If I study the map, plane and compass both go off, wing down, compass swinging wildly. If I watch the ground, I lose my heading. It is a confusing business, and I begin to sweat lightly. Honest to God, I don't know where I am, and I've been in the air barely a quarter of an hour.

Off to my left I can still see Lake Candlewood. It seems to be bearing about right, although it is a bit closer than I would have thought it should be. Below, villages and farms slumber in the noontime warmth of the balmy spring day. I see cows in a pasture, roads winding around through the low hills. But I see no lakes, no landmarks I can identify, and I am beginning to sweat rather in earnest now. If I saw an airport, I think I would go down—but what an embarrassment to have to ask: "Where am I?"

Suddenly, to my left, I see a fairly large lake coming up. Hastily I check my map—aha, there's one! It's right near Middlebury; in fact, the town is just beyond. Hope flares—it is also right smack on my course. The lake is to my left, Middlebury would be just to my right, my course would lie between the two. I strain my eyes ahead, watching this blessed blue body of water slowly coming toward me. There is a town beyond, not quite where it should be, but a town nonetheless, and my hopes soar.

They soar, but not for long. The closer I get, the more this place feels wrong. The town is not to the right of the lake, it is straight ahead and beyond it. To go between the two I would have to turn almost 90 degrees to the left, and this just doesn't seem to fit. I would have to be heading southeast to make the thing come out right, and even then the lake would be on the wrong side. Uneasiness comes up in me again—and then, as I get really close, I see this lake has an island.

It is a large island, very well defined. Lake and island both are plenty large enough to be on any map. But Middlebury's lake, study the map as I will, shows no island at all.

I look all over the map, trying to find a lake with an island somewhere near my course. There is a whole confusion of lakes and ponds coming up around Waterbury, but nothing with an island. I forget all about my heading as I search. This seems too important to worry about compass headings.

Now I am practically abreast of the lake, and I clearly see the town beyond. There is a railroad track; there are highways and a small river. There is no railroad track at Middlebury. I start sweeping the map in ever-widening circles—and then I find it. At Litchfield, miles north of where I think I am, is a lake, an island, a sizable town beyond, railroad track, highways, river—it all checks out. It checks out just fine—and I am fifteen miles north of where I thought I was, an infinity of distance on the map.

Now I have to work out something entirely new. While I fly the airplane along the lake toward Litchfield, I scan the map again. Thank God for the omniranges! There is an omni leg running right across between the lake and Litchfield to New Britain and the Hartford omni station. It is plainly marked: 112 degrees to Hartford. All I have to do now is turn right, get on a heading of 112 degrees and hold that until I reach New Britain, at which point I can turn left and head for Hartford, barely five miles away.

This, I say to myself, is improvised navigation of the highest order. I make my turn, pick up my heading, and away we go.

Now begins something which is still mysterious to me.

I am heading toward a river with a number of towns. As I pass over it, I see a railroad running alongside of it. Factory chimneys send streamers of smoke into the quiet air. It is a wide, sandy river, and the towns are pretty big. I am having difficulty holding my heading; things are jumping around. I am at about two thousand feet, and the sun-warmed earth is bothering me with thermal currents which bump and bounce. But things look okay—according to the map that must be Thomaston below, and Terryville beyond. Only they aren't really in the proper order. In fact, as I look it over, there are towns all over the place, and ahead of me another river valley is coming up.

Suddenly I remember. I have forgotten to add compass deviation. I check the little card which gives the correction for the compass in the plane—"For 120°," it says, "steer 150°." Thirty degrees of deviation—boy, that is a lot. I must be way north of where I think I am. So I turn right, heading southward.

Confusion now begins in earnest. I am now heading almost due south, and all I can see is sandy rivers and towns. Industrial Connecticut! Why the hell doesn't somebody mark these towns! I think bitterly of what a mess I am making of this first flight. I think apprehensively of sailing on and on into this peaceful day, winding up I know not where. I look yearningly for an airport somewhere below, but there isn't a one. I do everything, in fact, but really think; and this is an occasion where thinking is called for.

Suddenly, ahead and off to my right this time, but unmistakable, I see the same lake with the island. Litchfield! How did I get back here? Never mind, there it is; Litchfield has saved me again.

I circle the town very carefully, identifying everything with care. Okay. This *is* Litchfield. Take up your heading again, idiot, but this time take the right one and stick with it. No matter what happens, stick with it. And make that compass work for you, too.

Slowly I begin to think—and remember. How many times have I heard it: don't try to fly with eyes riveted to the compass. Get your heading, then spot a landmark up ahead, a hill, a lake, anything, and fly toward that. You can go crazy trying to make a compass stand still, and certainly a magnetic compass will drive you to an early berth in the bughouse. I fly the airplane carefully straight and level, pick my landmark and let the compass go.

Time, meanwhile, has stood still. I have no idea of how long I have been flying. That's another mistake—it means I have no idea how far I have been flying, or for that matter how far I must fly, or how long, before I do hit New Britain (if I ever do). I fumble in my map case for the protractor, and with shaking hands I measure the distance between Litchfield and New Britain. Twenty miles. I make a hasty calculation—if I am flying at 120 miles per hour ground speed, which I almost certainly am not (incidentally, did I ever check for wind drift? No), then I would get to New Britain in ten minutes. Call it fifteen. Thus I re-establish a rational relationship with time and the earth; I am no longer swimming helplessly in eternity, and already I feel better.

The countryside unreels below me. I pass my landmark, check my heading, pick another. I let the towns go by; it's hopeless to try to identify them. I pass over a river valley, and a curious landscape unfolds beyond. Lots of white flats, odd-shaped, like salt flats. The land is dotted with them. The horizon shimmers. Everything below is green fields with these curious splotches of white. The sky is a beautiful blue. The horizon shimmers—and then, in the shimmer, I see shapes. Clouds? Buildings? They must be big buildings. Suddenly I realize: Hartford!

It seems a miracle, a mirage. But there they are—Hartford's tall towers, the palaces of the insurance companies. Slowly they take on shape and form and solidity. I am slightly north of my intended track—wind!—and here I am. Brainard Field lies just beyond.

I feel a strong temptation to land. I have been flying for only forty-five minutes, but as far as I am concerned Brainard Field is Le Bourget after flying nonstop from New York; it is solid ground after an eternity in the air. It is now 12:30 as I pass over Brainard. All this stooging around that I did back there—getting lost and finding Litchfield and getting lost again—all this has taken only about ten minutes more than the normal flight time from Danbury to Hartford. It is a sobering revelation of how deceptive time in the air can be—time, and distance too. Up here in the empty blue, some things seem to take much longer than they do on the earth below; and by the same token, things actually happen much faster than they seem to. Among the many things that I forgot, for example, was that while at 2,500 feet I may seem to be drifting along, I am covering distance on the ground at anywhere from 90 to 115 miles per hour, which by earthly standards is pretty quick. Thus when I turned north in the beginning to fly upriver—my initial mistake—I might have flown anywhere from five to ten miles north, whereas actually I wasn't much more than two or three miles off course.

Also, since the river really ran not north but northwest, angling back from my intended course I was turning at an entirely different angle when I turned back off it again—far more to the northward than I meant to. Add a compass deviation of 30 degrees which I failed to take into account, and it is small wonder that I wound up in Litchfield, fifteen miles north of where I should have been.

But the biggest mistake of all was in totally forgetting what I had learned on the omni flight to Lock Haven with Bill Strohmeier: never try to get back on a course by turning sharply and cutting across it. Invariably and inevitably you will overfly it. Then you have to turn nearly 180 degrees to cut back on it again. The result is that you will proceed to your destination—if you proceed toward it at all—in a series of sharp zigzags back and forth along the course. You will never know whether you are anywhere near it because you will never get a steady compass reading; the compass will be forever swinging wildly, trying to keep up with the zigs and zags. The only way to correct a course error, omni or compass course, is to take a small cut at it—20 or 30 degrees—angling back *toward* it, not across it, correcting back and forth with smaller and smaller corrections until you are squared away. In this way, at worst you will be flying a series of shallow, very shallow S-curves along your track; at best you will soon be back on it again.

With the safety of Brainard below me, I now make all sorts of resolutions about things I will never do again. I take up a careful heading, pick a landmark, and fly with care. To my left, the river disappears northward almost at right angles.

A broad highway likewise disappears at about a 45-degree angle. Manchester lies ahead, parallel to the track I fly. I have twenty-five miles to go to Windham. There are lakes to watch for, a river to cross. One after another they appear, and I make damn sure that they are what they are. So I fly as precisely as I know how. I watch my wind drift, peering over the side of the airplane to see if the world is sliding by sidewise. It is, so I make a small correction. I hit everything with such exactitude that when I suddenly see an airport below I simply don't trust myself that this could be Windham. Without even thinking of the mike fright that has oppressed me in the past, I reach for the microphone, tune transmitter and receiver to Unicom, the private pilot's frequency on 122.8, and speak my first words as a solo pilot into the empty air:

"Windham Unicom, Windham Unicom, this is Tri-Pacer 7226-Delta. I believe I am approaching your field at 2,500 feet from the west. Would you give me a short count, please."

The reply comes with startling clarity. "Tri-Pacer 26-Delta, this is Windham Unicom. One, two, three, four, five, four, three, two, one. Over."

It is a woman's voice. While I tune in firmly on the count, I am still wondering. I am now passing just over the edge of the field. "Windham Unicom," I say, "I am not sure if I am over the correct field. I am now passing just north of the field. Do you have me in sight? Over."

A pause. Then: "Tri-Pacer 26-Delta, yes, I have you in sight."

"Two-six-Delta," I reply. "Roger, and thank you. What is your active runway, please?"

Back comes the answer. "Our active runway is two-two. An Air Force Navion is practicing on one of the long runways."

I roger and thank you and look overside as I start a gradual, descending right turn. Runway 22 is nice—long, smooth, right into the wind which blows streamers of blue smoke from a grass fire somewhere beyond. As I turn, I keep a watchful eye out for the Air Force Navion. Suddenly I see him below, a silver speck drifting onto another runway. I watch him land and take off again, circling away; then I am down at pattern height, into my downwind leg. I turn base and final; Runway 22 swims into my field of vision, comes closer, flattens out as I flare. I hardly feel any wind at all. We float for a couple of seconds as I hold 26-Delta off the ground; then, with a whisper of tires, we are down.

Windham seems the friendliest place in the world. The voice on the Unicom is buxom and cheerful, black-haired with bright blue eyes, a youthful matron sweeping out the office as I walk up. She signs my logbook, apologizes for having "confused" me with "that Navion—it's just that they don't have Unicom and I wanted to be sure one of you knew the other was around." The Navion's crew come in while I am there—a startlingly young Air Force cadet in immaculate uniform, an older staff sergeant. I work over my next leg with extreme care. By contrast to the way I prepared for Leg 1 of this trip, here is what I have noted down on a clean sheet of paper in a notebook:

LEG 2—Windham-Westerly
MC—158 DEV. 30 R CC 188 (*with the compass course circled for emphasis*)
DISTANCE 47 miles
NORWICH RADIO—109.8——— .—. .——338 FROM 158 TO
WESTERLY UNICOM 122.8

This is the information I need to have at my fingertips, and it is right here, on my little knee-desk, where it always should be, ready at hand. I take off at 1341 hours, feeling much better. I am on course at 1348, and I hit Westerly right on the nose, landing at 1413. I repeat the same process at Westerly, noting all headings, distances, radio frequencies and, after gassing up, take off again. The flight is beautiful down the coast; the air is crystal and millpond-smooth. At New Haven I switch tanks, take up my new heading, watch my check points, and land at Danbury at 1540 hours, a weary but much wiser man.

WEDNESDAY, MAY 6
This is the second day, and it, too, has something special. Joan Dickinson is coming up from Teterboro at one o'clock with the Piper company's Press Tri-Pacer, N-9013-Delta—the plane I soloed on at Martha's Vineyard nine long months ago (the gestation period of a flier, at least this one, is about the same as a human baby's, it seems). We have old 13-Delta for twenty-four entire hours—all this afternoon, all the next morning, and the next afternoon. It's almost like having our own airplane.

The trip this day is a long one, with many new problems.

Stan Kenecko set it up for me: first leg to Albany, 90 miles away; then from Albany to Brainard Field in Hartford, 88 miles across country that on the map looks mostly empty, wild hills with little or no landmarks; and then from Hartford to home, 50 miles. Albany is an airport of entry on the Canada run; it has a control tower, airlines, the works. It should be about two and a half hours' total flying time, possibly more; a real journey.

This time I have prepared things carefully, very carefully indeed. I have my headings, with space left to put in 13-Delta's compass deviation so I won't forget that again; I have radio frequencies and omniranges; I have distances; I have reminders to fly to the right on airway legs where I may be meeting airline traffic—I have everything. At the last moment I find I have marked my course to the wrong airport—Tri-Village instead of Albany Municipal, which is about three miles farther north in the bend of a river. This causes brief consternation, until I decide to make a virtue of the error—I can get to Tri-Village (one runway, turf), call the Albany tower from there, and thus have plenty of time to plan my approach and landing. On the way to the airport I also remember another important point: airport altitude. Behri looks it up and notes it: Albany altitude, 288 feet.

At 1350 I lift 13-Delta off into the beautiful, warm blue sky. At 3,000 feet we square away, I tune in Poughkeepsie omni (112.2) and we march. I am going to use omni in the background, so to speak, to check my pilotage.

In this flight plan I have even noted down a time check: Dover Plains. Now, as we head up along Lake Candlewood in rather bumpy air, I improvise a shorter one, figuring how long it takes me to reach a protruding arm of the lake, ten miles from the airport. Somehow or other, I make it five and a half minutes, based on an assumed ground speed of 110 mph. It works, quite well, and I am pleased. I predict myself an estimated time of arrival at Dover Plains, twenty miles out, and that works out quite well too.

The air is getting rough and rougher. We are flying over hilly country now, with mountains in the offing. I decide 5,000 feet might be a smoother altitude, so I advance the throttle to climb power, 2350 r.p.m., and up we go. It is somewhat smoother up there, but this is still the roughest flying I have done for some time. Perhaps 6,000 or 7,000 might be better still, but I am not yet prepared to fly that high, so I just ride it out.

Thirteen-Delta has a lot of instruments. She has, in fact, a full panel, with gyrocompass, artificial horizon, rate of climb indicator, manifold pressure gauge, turn and bank indicator and an Omnigator Mark II radio for navigation and communication, with a whole slew of channels. It works beautifully. We keep Poughkeepsie right in the middle of the little blue and yellow omni indicator just with check points. When we pass over the beacon, I watch the needle tremble, flutter back and forth, then switch while in the little window the red flag goes up and TO changes to FROM. Hot dog! This is living. But for the rough air, this would be a really dandy trip so far.

Landscapes from the air are wonderful. After Poughkeepsie omni, we pass the last of the mountains and ahead I can see a promised land, the smooth, green, slightly hazy plain of the broad Hudson River valley. It is a feeling really like climbing a mountain and standing on the peak to see the end of the wilderness and the beautiful man-kept land below. We sail across the last great brown ridge and the earth falls steeply away. The air smooths out and we are traveling across a vast and busy plain, remote yet bustling with criss-crossing highways, railroads, cities, villages, factories smoking, fields being plowed. Through the middle of it winds the wide river, its banks lined with towns, high hills beyond on the other side. Gradually we come upon it, and I start checking map with earth again. It is all there—the factory below with its mines (marked with crossed pickaxes), an airport to the left, islands in the river ahead and, far far ahead, barely visible in the warm haze, the bridge at Albany.

I am due to cross the Hudson River at a bend just above one elongated island and below two more. The river here looks almost small, split up, shallow and winding. A tug and a barge are etched against the brown-green stream, trailing a tiny feathered wake. Smoke trails from a mountainside on the steep opposite shore. Thirteen-Delta is humming sweetly. In the smooth air I have at last found my proper trim and air speed. I realize that all the way up, in an effort to hold my altitude in the bumpy air, I have had her trimmed too high. I have been mushing through the air at about 100 mph, wondering why she didn't go faster, bringing her up again when, thrown off by a bump, she would dip her nose and get up to 115 or 120. Now I have her trimmed straight and level and she shows an indicated, steady 110 mph at 2200 r.p.m.

I am on Albany omni now, 116.9. The range station is on the municipal airport, but I keep the needle slightly to the right so that I will hit Tri-Village a bit to the southward. My next check point is a beauty—a huge railway marshaling yard, which should be impossible to miss—and there it is, coming up ahead. Tri-Village airport should be just beyond, and I start looking.

Finding a turf field, I have learned, can be a real headache. Unless there is a hangar and some airplanes, they look just like any grassy field, particularly at this time of the year when the grass is just beginning to grow and the runway does not have to be kept mowed. Perhaps Tri-Village has no hangar. Certainly it doesn't seem to have any airplanes. Twice I think I see it, but it turns out to be just another farm. I must have passed it, and I make a mental note: turf fields are not good landmarks. There is no point in circling to find it; I turn off to the right and fly on to look for Albany.

Ahead I see the river, curling across my path. To my right is the city, and the Hudson bridges arching over the river. I tune in the Albany tower, and hear them directing traffic. There seems to be a good bit of it, the kind I've never yet had. They are clearing a Convair in now, holding a Constellation on the runway. And 13-Delta is coming in. It's time to let them know about Student Pilot Me.

"Albany tower, this is Tri-Pacer 9013-Delta. Over."

"Tri-Pacer 9013-Delta, this is Albany tower. Over."

My mouth feels a bit dry. Convairs. Constellations. Towers. Me. We are flying smoothly at 2,500 feet, but I don't see any sign of the traffic they are talking about. "Albany tower, this is 13-Delta, approaching the field from the southeast at 2,500 feet. Request landing instructions, please."

They give me clearance to enter the pattern, the runway number, the wind—light variable, from the east—altimeter setting, time. I roger and strain my eyes ahead. No planes, no airfield, nothing. I fly on in silence. I don't know what to say.

"Tri-Pacer 13-Delta, this is Albany tower." Oh-oh, they're after me now.

"Albany tower, 13-Delta."

"Thirteen-Delta, would you give us your position please." The voice is perfectly friendly, but I think of Convairs, Constellations. . . . Then, ahead, I see a field.

"Albany tower, 13-Delta here. I seem to be coming up on your field from the east, on my left. Do you have me in sight?"

I can see the field plainly now. It is a big one, concrete runways, taxi strips, tower, hangars. But no traffic at all.

"Thirteen-Delta, no, we do not have you in sight. Where are you?"

"Albany tower, I am now circling this airfield. I am right over it. . . ." I really don't know what to say; I feel apologetic, apprehensive, and confused.

"Thirteen-Delta." The voice is still perfectly friendly. "You may be over Schenectady. Advise trying to call them to see if they have you in sight."

Hastily I check Schenectady on the map, tune in their frequency, call. They come back promptly. No, they do not have me in sight.

Albany comes back. "Tri-Pacer 13-Delta. Sir, are you on a solo cross-country?"

With an immense feeling of relief, I reply: "Albany tower. Yes, sir, I am." I know now they understand.

"Thirteen-Delta, advise us when you have located yourself."

"Roger, Albany tower, I will do that."

Okay, Buster, here you are. You are at 2,500 feet, circling an airport somewhere near Albany. For Pete's sake, calm down. You should be able to see where you are and figure out where you must go.

Circling slowly, I check the field below with care. There is absolutely no movement on it at all. A large silver plane is parked near one of the runways. A whole fleet of silver, swept-winged jets is parked on a ramp before a hangar. The airfield is on the far side of the river, just above a bend. The city of Albany is to my right. No other field is in sight.

Now, just as carefully, I check the map. There are half a dozen airfields around here. There is Albany. It is on the south side of the river, the opposite shore from this field. It is on a bend that points south, directly toward it. My field is on a bend that points north, directly toward *it*. It must be this field here. All I have to do is turn east, toward Albany, and follow the river. I've got to hit the field if I do that; I can't miss it.

I turn away from the silent field below and follow the river along. In a couple of minutes I see an airfield ahead. I see a silver airliner just taking off, heading south. I see another taxiing. I am here.

"Albany tower, this is Tri-Pacer 13-Delta."

"Thirteen-Delta from Albany tower. Please stand by."

I pass north of the field and start circling slowly to the right, keeping my eyes well peeled. The airliner below has taxied to the runway and paused. He is probably running up his engines. Now he begins to move, gathers speed, takes to the air and, climbing, begins a slow turn to the left. I watch him carefully. He is far away, but I am turning toward him, and I am conscious of the fact that as I lose altitude I will be approaching his level. Now he straightens out and disappears.

"Tri-Pacer 13-Delta from Albany tower. We have you in sight now. You are cleared to land. You are number one to land." They give me the runway number, wind, and altimeter setting as I go into my downwind leg.

Only as I am taxiing toward the ramp do I realize that the runway lights are on, turned on to guide a confused student pilot back to earth on the first big airport he has ever known.

Coming back to earth at Albany was like returning from another world. As I parked the airplane and climbed out, the solid ground felt strange. The line boy who had waved me to my place in the line of airplanes came up to the door; I looked at him as though he were a man on Mars. I had been somewhere far, far away, alone in the sky with my airplane and my problem, and had been utterly absorbed. It is a quality of flying that is unforgettable and unique in human experience, a state of mind induced by this abrupt severing of the umbilical cord that ties man to his earthbound existence which is an act of commitment like none other.

When the airplane is committed to its take-off, the pilot says good-by, for however brief a time, to all the human conditions which man knows on the ground. Even the sea does not produce such a moment—leaving port, the sailor turns his eyes ahead to wind and waves and the uncertainty of his voyage, but the land lingers a long time in his sight; there will be hours still when he can see and smell it and turn back. But the pilot is committed to the air when his wheels leave the ground. There is no turning back for him once his wings have felt the surging lift of their own element: he must go on, he must climb and fly before the he can land again.

And he is at once remote, a lonely and a dedicated man. As he depends upon the airplane, so the airplane depends on him. The bond between them, between the living, thinking, feeling being and this mechancial miracle which he helped to create and fly (for all mankind has shared, to however infinitesimal a degree, in the creation of the things which man has built)—this bond is deep. It goes below the level of consciousness; it makes the man a part of his machine; the two are one, the pilot acting through his plane's controls, unthinkingly, by pure instinct, concentrating with some deep, deep part of him that in normal life is seldom called to wakefulness, on the problem of bringing his ship home.

This is what I felt at Albany as I stared about me at the warm and sunlit earth, listening unhearingly to the sounds of busy men, seeing unseeingly the familiar bustlings of my land, my country, my fellow human beings. I had been far, far away, 2,500 feet above this land but a million miles remote, alone, alone responsible for my life, my only tenuous tie the friendly voice of an unknown man, the helpful act of an unknown hand that turned on the lights for me. I have never loved man more, nor life, nor flying, which alone has wakened that deep part of me.

THURSDAY, MAY 7

This is the third day, the weather day. Yesterday's flight from Albany to Hartford and home was completed without incident. Today, at eight-thirty in the morning, I am to go to Westfield Field in Westfield, Mass., thence to Groton, across the river from New London, and home. This is to be an omni flight all the way.

The day is cool and cloudy as I drive to the airport—high clouds with broken patches of vague blue sky and veils of haze in the valleys. The airport is quiet when I arrive, the day's activity only just beginning. The air is dead still. My airplane is standing on the ramp. Only Ken, the line boy, and I are there, and together we gas up 13-Delta and carefully check her out. "Keep an eye on the weather," says Ken as I climb in. I nod, say good-by, taxi out, and take off on the west-east runway, climbing above the hidden valleys and the still sleepy town.

I circle once, level off at 1,500 feet and pick up the Wilton omni. The chirping musical notes, ending on an easily recognizable dash, are warm and reassuring in the gray, smoky sky. I want to find the range station, which is only a few miles away, so that I will henceforth always know where it is in relation to my home airport. I fly carefully, keeping the needle centered as long as possible—a delicate job as the beam gets narrower and narrower near the antenna which is its source. At last it swings all the way over; the red flag comes up, TO changes waveringly to FROM, I tip my wing and mark the place, in the wooded hills just west of Georgetown. Then, keeping a promise, I swing off and head toward the village of Wilton, descending slowly to 1,000 feet. I want to circle my sister's house and give a waggle of wings to my small nephew, Stephen. I find the house, a toy house in a tufted landscape of toy tufted trees, circle twice and then take a run over it. Below I see small figures waving madly. I waggle to them, smiling, wishing I could call hello. Then I head out over the omnirange, pick up my heading, center the needle, and fly.

It is a strange day, quite unlike any I have ever known before in the air. The world is silver-gray with a blue sheen; the clouds to the east, where the blue occasionally peeps through, have a golden touch to them from the hidden sun. The water of Long Island Sound off to my right is a great shimmer of silver, streaked with light. At 1,000 feet the earth seems close below me, but all of it is touched with silver too, and the haze lies like spun silver glass between the hills.

Slowly I climb out until I reach 2,500 feet. I pass Danbury on my left, reach Newtown, see the familiar hospital, the bridge at Sandy Hook, the great sandy gash of the new highway. And then I begin to wonder.

The sky ahead is dark, really dark. There is no earthly horizon, only a horizon of cloud, high up; and from this darkling line a curtain of haze hangs down, laced with dark streamers that look like rain. The more I fly toward it, the more it looks like the kind of weather I have so often been warned against—the soft fog that seems ever to retreat before you until suddenly you are in it, closed all about with clammy opaque wetness, all sense of direction, all relationship with the earth gone. It is in such weather that the inexperienced pilot, not yet trusting his instruments but fearfully conscious of his own frightened senses, loses his equilibrium and goes into the so-called dead man's spiral. I have already known the feeling, flying with Bill Strohmeier in the Comanche in a hazy sky, concentrating on instruments alone. It begins with a sense of having one wing down; the instruments say no, you are flying level, but the feeling becomes overwhelming and, too often, the pilot will obey his instinct and bring a wing up. Now he is really flying one wing down, on the opposite side, and the plane begins to turn. As it turns, its nose drops and it picks up speed. The pilot, now acutely aware only of his own feelings of things going wrong, pulls the nose up. The ship loses speed and falls off still more. A fatal sequence of wrong corrections, overcorrections, and panic action has begun: the spiral steepens, the plane descends, the pilot pulls back on his wheel and fights the rudder and that's the end. "Observers saw the airplane come out of the clouds in a steep, spiraling dive or spin. . . ." So many accident reports have ended that way.

I make up my mind. I don't like that weather ahead. I have no pressing business in Westfield; I will never have business so pressing that I will fly into weather I don't trust, or in which I don't trust myself. I turn around and go home.

Back at the airport the decision is universally approved. We discuss alternate routes, and Russ finally suggests I go to Hartford and from there to Groton and home. This route is a little south of the weather I saw. "You can feel it out," he says. "If you don't like it, just turn back."

I take off and head for Hartford. The darkness has lightened somewhat; the curtain of haze has thinned. I poke into it cautiously at 1,500 feet. I can see the ground clearly, and as long as I can see the ground for some distance ahead of me I know I am safe, with time and room enough to turn if it should thicken. But the sun burns it off, and I reach Hartford with no trouble, landing in the full heat of a humid morning.

If I had had pressing business in Springfield, I could have been there less than an hour late—and alive.

Back in Danbury that afternoon both heat and humidity have become oppressive and thunderstorms are clearly in the air. There is still almost no wind, but a front is expected to come through some time in the evening. Thirteen-Delta has to be back at Teterboro by the end of the day, but there is still time for a short trip for me. At a quarter of two I take off for Dutchess County airport near Poughkeepsie, twenty-eight miles away to the westward. From there I can fly back to Bridgeport, a trip of about fifty miles, passing right over the Danbury airport. It is a good little trip, all in the local area so that I can get home quickly in case of sudden bad weather, and it will give me just enough time to complete my required ten hours of solo cross-country.

What wind there is has shifted south, so I take off toward the valley, climbing between the ridges of the brown hills on which the green is just beginning to show. This is always a dramatic take-off, for on either side the hills are close; there is a sense of intimate relationship with both the air and the earth, a real feeling of finally, exultantly, pulling free to turn and climb into the blue over the defeated heights. As I climb out I scan the sky all around me—a haze-hung, heated, waiting sky with dim cloud banks, some barely formed, others assuming shape and solidity, but all exuding that sense of heaviness, of waiting for something big, inevitable, and violent to break their spell, release their weight, and clear the burdened air.

As I reach toward 2,500 feet the earth grows dim below. The haze is thickening; better to take the bumps and keep the ground in sight. I want to make this flight by pure pilotage, by following the ground as I feel my way through the air. I am beginning now to feel consciously a sense of mastery and brotherhood with this element and this machine, to taste and enjoy and exert the bond that was welded between me and my airplane at Albany. I am beginning, at long last, to fully trust myself and Mr. Piper's product, to feel the strength of knowledge, the strong supporting background of experience that tells a man what he can do. I revel in it as I fly along, singing along with the engine, smoothing the bumps with hands that know now without having to be told what they must do.

The check points come up, minute by minute, one by one: Brewster goes by, off to my left, and Carmel and the small lake embraced by twin railroad tracks, black lines on a rusty land. Hills rise ahead of me, and I climb. I soar over the top of them, they drop sharply away and ahead of me I see, with sudden clarity, a wall of cloud.

This is really thick weather. It climbs into the sky as far as I can see, a great hanging darkness ominous with wind and rain. It is so well defined that for a moment I think of simply flying around it, past it, and to the Hudson beyond to see whether Poughkeepsie is in the clear. But even as I turn I realize that if I should do this, I would cut off my retreat. The storm might well be moving right toward Danbury; it would then be between me and home.

And as I think this, the wind hits me.

I feel it through my body as plainly as I would have felt it striking me abruptly in the face. My right wing rocks and jerks spasmodically into the air. I have started turning back, away from the storm, and the wind is pushing in great gusts under my uplifted, turning wing. "Shallow on the upwind side, steep on the downwind"—the long-ago advice from Gene Brandon flashes through my mind. Already my hand on the wheel is steadying the plane, bringing the wing down as we are swept back toward the hills. Holding her as steady as I can, I complete the bucking, bouncing turn. This, I tell myself, is going to be a rough ride back. I have the feeling of being chased now, of having to run from something; I feel I can hear the wind whistling and growling at me malevolently from that high, dark cloud.

As yet there is no danger, and I know precisely where I am. I consider carefully. I can certainly get back to Danbury, but that would make twice today that I have turned back. I have no pressing business in Poughkeepsie, of course, and anyway it is out of my reach now, but I don't want to abort this trip yet, either. It is not a matter of pride, it is simple reluctance to end a flight in which for the first time I am consciously feeling my newly acquired skills, tasting and enjoying the just discovered bond of confidence between myself and my plane. I don't want to end the three days on a down-beat note; I don't want to simply turn and run.

An airport lies directly ahead of me—Stormville. I have flown over the place, and have long wanted to go there, but Stormville is small, with a dirt field, and I have not landed on dirt or turf since Martha's Vineyard. Now I see my chance, and I seize it. I feel I am ready for an emergency landing, and I can always excuse this one as an emergency on the grounds of the coming storm.

Low down, at less than a thousand feet, I overfly the field. I see the wind sock, stiff and straight, swinging with the shifting, gusty wind. There are two runways; the wind is blowing right between them. I choose the longer one and begin the circle into my downwind leg.

I am low now, at 800 feet, and 13-Delta is bucking and bouncing like a bronco in the threatening sky. But my hands and feet know what to do. I add a little power; the engine answers at once, and the nose lifts. The wind, sometimes cross and sometimes partly behind me, blows us along. I lift the nose a bit to cut my air speed, watch the hand on the dial jerk spasmodically back to 100, to 95. I put on first flaps—and then I remember the prison.

What this prison is I do not know, but it lies directly beyond Stormville airport and I am heading straight for it. There is a stringent regulation against flying over prisons, and I would certainly be violating it by flying over one this low. There is nothing to do but turn away, low as I am, bouncing around as I am, and make a dogleg around it.

This is something truly unscheduled, and my heart is in my mouth as I carefully turn the plane. I cannot let my flaps go; I am too low. I must be careful with my air speed; 100 mph is the limit for flying with flaps down. I nurse the throttle forward, keep the nose up, nurse my bank, and feel a stab of pride at the way my ship reacts to my instinctive movements. We sweep past the prison walls, I get a brief, unforgettable glimpse of grim towers, boxlike buildings, a tall brick chimney, and a gray-brown yard with immobile figures standing. Then I am past, and must turn again, into base leg and final.

Leaping around though we are in the sky, our flight is controlled and we are descending slowly. I come around on my final approach and see the runway jumping crazily before my nose. Wind swings me back and forth. I am high, and in this wind my drift and float will be strong. I must land short; I don't have three quarters of a mile of well-paved runway with wide grass borders to land on here. I put on full flaps and close my throttle all the way, pushing the wheel forward hard to hold the nose down against the gusty force that lifts me.

The air speed drops to 90, 85. I ease back on the wheel. The ship hovers at 80 miles per hour, slow but safe with full flaps and this wind. The tail slews as a gust strikes; my feet push opposite rudder in the same instant. The wind is hitting me squarely on the side now, directly cross, but I am committed to my landing. I hold the wing down into the wind, correcting for my sidewise drift, easing it up and down as the gusts buffet me.

Down, down. The end of the runway flashes past below. Too fast. I ease the wheel back. Too high; I cannot flare out yet. A gust hits me; I hold the wing down, ease it back. My flight is straight and true; we are not drifting. Now the runway flattens suddenly as I drop in a momentary easing of the wind. Back wheel! We float for just an instant, just long enough for that baleful wind to strike me one more blow, for me to strike back one more time with lowered wing. Then we hit the ground. We bounce once, slightly; I shove the wheel forward, and we are down, rumbling fast over loose clods of dirt, raising a streamer of dust behind.

Shakily I park my 13-Delta, climb out and walk up to the house. I have a quick impression of a lovely, old-world setting. Stormville, my haven! Of all the names! There are giant trees in the front yard, comfortable reclining chairs. A man is sitting on a rocker on the front porch; he raises a hand in greeting as I advance. A sense of peace steals over me, of peace and happiness. I am down; let the wind blow, we have made it, my airplane and I.

I will never forget Stormville, and I hope often to visit there again. I stay for perhaps half an hour, waiting to see what the weather will do. I sit and chat with Stormville's people; I get my logbook signed, we exchange names in an easy and comfortable, old-world way. I learn a little bit about the place—small as it is, it is a busy little field, particularly for week-end fliers, and I can well understand why. Inside the house the rooms are cool and large and shady; there are big, soft chairs, books and magazines, a snack bar and the feeling of long, lazy hours. Outside, a dozen planes are lined up in a neat row, well taken care of, ready to fly when the mood suits. No tower, no radio, just flying, as it was in the old, early days when Stormville first began.

I leave with real regret, but it is time to go. The storm is slowly moving to the northwestward; the air to Danbury is clear. I shake hands, say good-by, climb in, start my engine, and slowly taxi away. The wind has died. "You climb out over that hill there, past the fire tower," they have told me, "and you can practically see Danbury." I look out ahead into the wide, wide sky, advance my throttle and roll. A few drops of rain spatter my windshield as 13-Delta leaves the ground; I sing as we climb, a song of joy and exultation born of confidence and knowledge and we soar over the hills toward home.

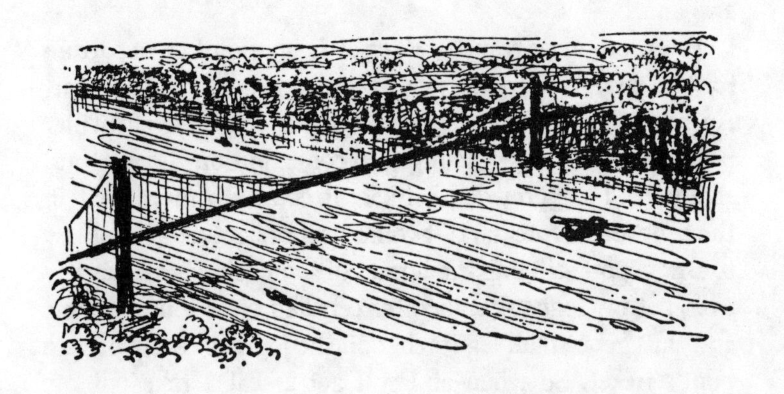

6. *How Not to Fly*

There comes a time in every student pilot's life when, for one reason or another, he forgets all the rules, throws away everything he has learned, and survives by the grace of God and the instinct which the hours and hours of drill in flying the pattern have bred into him. My moment comes on a warm summer evening after a sweltering day, when the upper air invites me with the promise of cool blue skies to escape from the hazy heat below and fly, just fly. Which is exactly what I do, with consequences which in due course replace the flush of summer heat with the burning blush of shame upon my cheeks, and make me want to apologize to my airplane.

Teterboro airport in New Jersey is my point of departure for this trip, and I reach it via bus from the New York Port Authority Terminal on Forty-first Street. I have no particular program in mind for the flight to come, but on the way out, standing in the crowded vehicle, it comes to mind that in the time left before dark—two hours more or less—I could fly to Bridgeport, circling the houses of some friends on the way over, get my logbook signed for certified cross-country time, and thus get a small leg further on the journey to my private pilot's ticket. So when at last I get a seat I take out my Sectional Chart, work up a few compass headings, make a rough calculation as to mileage and flight time, and consider myself prepared.

This is a mistake. Hasty planning always is, and even in the best of plans there are always factors which can unexpectedly throw things out of whack. Such as, for instance, that the airplane may not be ready to go.

I have counted on departing by 6 P.M. At ten minutes before six I check the plane, Tri-Pacer N-5927-Delta, and find the tanks are only a quarter full. The tank truck is busy somewhere else; it is six o'clock before it gets to me. By the time both tanks are filled and the truck has pulled away, it is almost 6:10. By 6:15 I have taxied out to the runway, checked magnetos, carburetor heat, gauges, controls and flaps, and reported Two-seven-Delta ready for take-off. Behind me, a DC-3 growls impatiently; ahead of me a Cessna waits, and down the runway a seemingly unending procession of airplanes comes in to land. At 6:25 I hear at last the tower's welcome "Two-seven-Delta, cleared for take-off." I am nearly a half-hour behind my improvised schedule as I soar off into the blue.

At 1,000 feet, having made my pattern turns into the downwind leg, I speak into the mike: "Two-seven-Delta, leaving the pattern," and head out in a long, straight northeastward climb. Off to my right, New York City bakes in its heat haze. The George Washington Bridge casts a silvery shadow across the shimmering Hudson River. I tune out Teterboro tower, tune in Wilton's omni, set my course on the 64° radial, and fly.

At 3,000 feet it is cool, peaceful, and wonderful in my little cabin. Every inch the master pilot, I throttle back to cruising speed, trim ship and lean out the mixture for maximum fuel economy. My Sectional Chart is laid out on the seat beside me for easy reference. In ten minutes or so I should be passing between Westchester County airport, a large and easily visible field, and Armonk, where my friend Ken MacLeish of *Life* is probably flying his Cub Cruiser. Feeling well taken care of and relaxed, I lean back in my seat, light a cigarette, and, one hand resting lightly on the wheel, stare pensively out of the window while I contemplate the joys of aviation.

On a rough calculation I have figured that Bridgeport, forty-seven miles away, should take me about half an hour of flying. Slightly off course is my sister's house in Wilton, which I would like to circle; add maybe five minutes for that. Then to Redding, to circle the house of a friend—call that another five minutes. Forty minutes altogether—okay, Bridgeport at 7:05. For the first time, as I review my figures, a small, disturbing doubt creeps through my mind: if I am to land at Bridgeport, get my logbook signed and fly back to Teterboro again . . . What time does the sun go down?

The nagging question disrupts my superficial peace of mind. Nothing outwardly has changed—I am flying straight and true, trimmed out and with the engine humming at 2,150 r.p.m. and 110 miles per hour air speed—but a chill now hangs in the air. Or rather, it is in the pit of my stomach, which asks quietly: Student pilot Knauth, have you planned this flight well?

I thrust the thought aside. After all, I am over familiar country. At this very moment I am passing over the runways and the hangars of the Westchester County field, shining in the slanting rays of the late sun. Wilton's omni signal is steady in my ears; I am dead on course. But I do notice that in order to hold 64 degrees I have to fly about 50 degrees; I have not reckoned with a wind which, blowing west by north, is drifting me off to the right. And with this thought, the moral temperature in my smug little cabin abruptly drops another ten degrees. Student pilot Knauth, coming back from Bridgeport you will be bucking a head wind of maybe 15, maybe over 20 miles per hour. Have you planned for this?

The answer, of course, is no—I haven't planned for anything. But this is not an answer that I give clearance to. Instead I close my mind to it and concentrate on trying to pick out fields I could get into in case my engine fails.

This is a never-ending and superlatively useful pastime. It may never happen; the chances against it are more than comfortingly large in this day of superbly crafted internal combustion engines, but it is an invariably interesting antidote for occasional boredom in the skies. I look at golf courses, pastures, plowed fields, and estate lawns, mentally picturing a Tri-Pacer coming down there, figuring wind and approach patterns as the minutes tick by. Every now and then I correct my heading, keeping the omni needle centered. Ponds and lakes swim past below me and, at length, I spot one that is familiar: the Norwalk reservoir in Wilton, near which my sister lives.

Now I am all business as I scan the tiny, scattered houses below. It is difficult to relate them to the familiar view one has from the ground—a road that turns off here, then makes a right turn there, sweeps left, merges into another; ah, but there it is! Suddenly it all falls into place: I see the house, the driveway, the garage, and I put on my carburetor heat, retard my throttle and begin a sweeping downward spiral. I put my wingtip right smack on the yellow driveway and hold it there, feeling pride in my pylon turn as I unerringly descend. But there is not a sign of life, no figures running out and waving, nothing at all. I suddenly remember—I am almost half an hour behind schedule and at this time she and the children are off in New Canaan, meeting my brother-in-law's train.

Ah, planning, planning, careful planning! What price planning now! Good God, it is nearly seven o'clock, I must be ten minutes away from Bridgeport at least—what time does the sun go down?

Two-seven-Delta gets her engine gunned, her nose trimmed up to climb. The knot in my stomach tightens. Brother, you had better move if you're going to get back to base tonight!

I climb and head out—where? I haven't laid a course for this contingency; I was going to head out peacefully to the Wilton omni station, then turn and take a heading of around 130 degrees from there to Bridgeport. Not that I really felt I needed a heading—after all, this was home country, where I lived. But now, all of a sudden, everything looks different. Haze obscures the skies ahead, and the ground below, while entirely visible, is strange.

This is the time when everything I have learned—everything but the instinct which has been drilled into me—begins to fade away, chased by the unreasoning process of fear.

There is no emergency, none whatsoever, but for the first time I begin to sense the implacable machinery of time running out. The sun is still high in the western sky; I can certainly reach Bridgeport even if I have to stooge around in the sky for a while to get my bearings and familiarize my senses with the land below. But reaching Bridgeport is not enough; I want to get back to Teterboro. And I begin to realize that from the very start I have not allowed myself any margin; that if I get back at all, I will be cutting it very fine.

First of all, I have to find myself. Land-trained senses take over; what I have learned in the air goes out the window. Instead of looking at my map, taking off from the known point of my sister's house in Wilton hard by the well-marked reservoir, I try to find myself with my eyes on the ground alone.

How fast an airplane flies! Wind and all, I am probably doing nearly 100 miles per hour; nor am I paying any attention to where I am flying in the sky; by the time I have got myself oriented, I am halfway across the visible landscape to Redding and my home in the hills. Only here do I find myself for certain. I also collect my wits a little bit. I look at my compass, for one thing—I am flying almost due east. At this point, to reach Bridgeport, I should fly almost due south. I turn and do so.

I also look at my map. A pattern of three reservoirs forms a triangle between me and the city; if I split this down the middle, I will hit the airport on the nose. Feeling a little calmer now, I head down toward the shimmering haze that obscures Long Island Sound, see the city come up, its buildings and bridges, and the point where the airport lies.

The point, the point—which point? I have never flown into Bridgeport before. I realize with increasing, cold clarity what an utter damn fool I have been. The knot of fear tightens again. I can't afford to lose time now, not if I want to get back. . . .

I check the map carefully once more. There is one point that sticks out farther than all the rest, and the airport lies on it. It *must* be dead ahead, there in the haze. I force myself to fly on, fighting the temptation to turn and search. I must be right—and I *must* act on my conviction.

There it is! Shining through the veil of smoke and heat the runways appear, straight and wide. My heart leaps to them—I reach for the radio with a hand that trembles and tune in the Bridgeport tower. I can hear them talking to another airplane. "Bridgeport tower," I say, "this is Tri-Pacer 9527-Delta, two miles north of the field at 2,500 feet. Over."

For a moment they chatter on. "Flight Six, cleared for take-off," I hear, and then Flight Six's answer. Far below, a silver twin-engined plane moves down the runway. Air-borne, he calls in again. "Okay, Flight Six" comes the tower's reply. "Next time we'll expect to see you in a Boeing 707. So long, now!"

"Bridgeport tower," I quaver again. "Tri-Pacer 9527-Delta, two miles north at 2,500 feet. Over." And wait breathlessly for their reply.

"Two-seven-Delta," comes the answer, loud and clear, "report on your downwind leg. Runway something-four"—I don't get this quite clearly—"wind southwest ten, altimeter three zero zero point eight." I am over the airport now and can see the runways clearly. There is a runway 24 almost at right angles to the one the airliner took off from, but it seemed to me the tower had said 34. "Two-seven-Delta," I reply, "say again your active runway, please?"

"Two-seven-Delta, that is Runway Two four," comes the answer. "Roger." I reply. I am heading almost directly across it. I am still at 2,500 feet. Happy as I am to have finally reached Bridgeport, I haven't even considered an approach pattern. Now how in hell do I get down to Runway 24 from that altitude, fitting smoothly into my downwind leg?

Ah, planning, planning! What price planning now? I start a wide circle to the right, then suddenly remember that airliner. Where is he, for God's sake? Did he fly straight out, or is he turning under me? I realize abruptly that I am spiraling down directly over the field. Suddenly the sky seems much too small for me and that silver-winged, two-engine job. What should I do? In a near panic I head out to sea.

Throttled back, I am descending now in a wide, fast turn. Instinct takes over; while my eyes search the skies, my hands automatically trim the plane back to an 80-mile-per-hour glide. Somehow, unrealizingly, I have put on carburetor heat; that's taken care of. I reason that the airliner must be long since gone. The thought calms me. I complete my turn and come into my downwind leg. I am at 800 feet now. I put on power to hold me at that altitude, report to the tower, and watch as the unfamiliar runways slide past.

Time to turn base now. I am still at 800 feet. I see the runway out of the corner of my eye; time to turn on final. The field swims into view, close below me. I am way high, still under power. I cut my throttle, pull it back as far as it will go. I float in at 80 miles per hour, and the end of the runway slips past below. High, high, I am way high!

Right now I am as rattled as I have ever been in my life. Altitude, the disappearing runway, the passing of time, my own foolishness and lack of planning—everything crowds in on me at once. I've got to get down! How? All of a sudden the final folly rings like a peal of salvation in my ears; put on flaps.

I reach down and pull the handle to first position. The airplane, already trimmed nose-high to an 80 miles per hour glide, balloons sickeningly. I push the stick forward and, in utter unreason, pull on full flaps. Up she goes even more, and my eye just catches the air speed, dropping from 80 to 60, and lower still.

This is no longer merely bizarre; this could be critical. I am still at least 50 feet high, a long drop if I stall. The falling air speed galvanizes me—I shove the throttle forward. The plane leaps, rises under the motor's pull. It never occurs to me to take some fast turns off my nose-high trim; but thank God, I do remember what Steve Gentle once said about putting on flaps: "Once they're on, boy, you're committed. Take 'em off, and the bottom will really drop out!" All I can think to do is get the airplane on the ground—with sheer, brute strength I shove the wheel forward against the combined force of her upward trim, the lifting flaps, the surging engine. And down she goes at last; the runway flattens out before me, I cut the power and, still idiotically pushing the wheel forward, set her down on solid earth.

The sweat is rolling off me as I come to a halt and turn. The full and final realization of what an utter fool I have been hits me and I am overwhelmed by a sense of shame. Talk about forgiving airplanes! I know what the phrase means now. What have I ever learned that could make me pull a chain reaction of stupidities like that!

I only hope, as I taxi to the tower, that no one saw what I was doing. It's enough for the moment that I know what I did. I take my logbook in and get it signed. "How much time left before sunset?" I ask casually. At least I hope I ask it casually; I feel anything but casual about what the answer may be.

"Well," says the tower man, "say about forty-five minutes. Where are you going?"

"Teterboro," say I.

He purses his lips. "Guess you can make it all right," he tells me, "but I wouldn't waste any time if I were you."

I don't. I am out of there and into the Tri-Pacer in nothing flat. I start the engine, taxi out, check mags and carburetor heat—and sit. A Cessna is in the pattern, turning on base leg. At last he comes in, touches gently, takes off again. Finally I am cleared. My forty-five minutes have shrunk to less than forty before I am even in the air.

There was, of course, no time at Bridgeport to lay out a course for Teterboro. I am squarely on my own now, on pilotage. But I am resolved to do this leg right. I learned a lesson on that first leg that I will never forget. But my troubles are far from over—I am up against something as dangerous as my own stupidity now, something utterly implacable: time.

It is there before me in the setting sun. As I climb out, I find myself flying directly into it. It still stands nearly 20 degrees above the horizon, and it blazes full in my face as I square the Tri-Pacer away and head across Connecticut toward the Hudson River and New Jersey.

The sun seems to fill my entire field of vision. It is bright, hot, a burning yellow-gold. It streams in through the windshield, blotting out the sky. I am still climbing, and I concentrate on my instruments, watching my rate of climb, my air speed, my artificial horizon.

I sense that I am turning. Instinctively I look out ahead to orient myself between the earth and sky. There is nothing there but incandescent light, shining on a bank of shimmering haze. There is no horizon. I have no sense of up or down, and for one panicky instant, I hover on the edge of vertigo.

Once again, however, training saves me. I look out the side window, where the sun does not stab at me; there, faint in the haze but still visible, lies the shore line and Long Island Sound. I check my artificial horizon; it shows I am flying straight, wings level, holding course. I hold her until I reach 3,000 feet, then level off. Shielding my eyes with my hand, I find that I can fly into the sun and still preserve my equilibrium; the panicky vertigo is gone.

That problem licked, I am immediately faced by another. It is imperative that I return to Teterboro by the most direct route available; the very best I can produce. I must rely on my knowledge of the general terrain and split the distance down to the shortest that I can. The sun is sinking; I don't have too much time.

I check the chart quickly to bolster my memory of the route I have to fly. If I take a shallow angle away from the coast, I should hit the Hudson just about at the Tappan Zee Bridge. From there, I can follow the river down until I see the George Washington Bridge at the northern end of New York City—a right turn there, heading westward, will bring me straight to Teterboro.

The Hudson is one thing I can't miss. No matter what I do, I'm bound to cross it somewhere. That thought makes me feel more secure; I bend my efforts now to hit it at the point I want to.

Time passes, the Tri-Pacer drones on. Wilton passes below me; I see it, recognize it, but can only think of how far I still have to go. The sun is changing from yellow to orange, drowning slowly in the haze ahead. How long before it disappears entirely?

I recall a flight a few weeks earlier, when I put down at Westchester County field because the skies at Teterboro were so thick with haze that I feared I could not land there. I phoned to find out what the weather was. "Tell him to fly back at 2,000 feet," they said, "and he'll be okay." I did and I was; and now the memory helps me. I am at 3,000, flying at 110 miles per hour indicated air speed; I might speed up the airplane if I nose her down, increase the throttle just a bit, and aim to hit the Hudson a thousand feet lower than I am now.

I inch the power forward to 2,250 r.p.m. I trim the nose down just a trifle, until I see the altimeter needle start to fall. The air speed builds—115, 120, 125 miles per hour. We are descending very slowly, but we are covering ground; I have the feeling I have gained a little in my race with time.

Fields and woodland pass below me; then, ahead, I see Westchester County field. The sun shines red on its hangars; it looks very comfortable and secure. Ahead, by contrast, the haze is thickening. I look at my watch; already it is 7:40. How much daylight is still left?

Westchester County suddenly seems like an awfully sane and sensible place to go. The temptation that it offers is almost overwhelming. To hell with the sun, to hell with the passing of time! I can put down there now, and all these problems will be behind me. . . .

But if I do put down there, then what? Failure, complete failure. A taxi to the station, a train to New York; a train back in the wee hours of the next morning to get the plane home again. All this would be thoroughly justified in the face of an untenable situation. But is it justified now?

Carefully, I try to measure time and distance. By now it is almost 7:45. If I fly right, I should be over Teterboro in fifteen minutes. In fifteen minutes it will not, cannot yet be dark. All I have to do is to fly right. With what I have, I've got to plan, and make the plan come out.

So I let Westchester County pass. I know that if I fly the same course that I have been holding, I will cross the Hudson at the Tappan Zee Bridge. I can cut my distance right now if I angle slightly southward. Carefully, I take off ten degrees from my heading and stare ahead, into the still-burning sun, into the gathering haze.

Gradually I make out a ribbon of silver. At first it seems like just another lake or reservoir of the many I have passed. Then I see another below; I realize that the two are connected. The river emerges from its smoky veil—I have hit it just where I want to. Off to my right, the Tappan Zee Bridge snakes across it; now I can cut south some more, straightening my line of flight yet once again. The sun is entering the upper layer of haze. I have one final stab about Westchester County airport—should I turn back and be completely safe? No. I am convinced that I have time still to make Teterboro. If there were still confusion in my mind, I would go back; but now at last I am thinking clearly, and I am morally certain I can make it. And I go on.

Never have I flown an airplane more carefully. I am still nursing the Tri-Pacer downward, getting every bit of speed I can without sacrificing a correct approach. I want to reach Teterboro at 2,000 feet on the nose, and I don't want to have to climb back to that altitude because of having gone down too far. And I am making it. We are sliding downward on a long slant, using power and gravity to maximum efficiency. And now, to my left as I cross the river at last, I see the towers and the blinking beacon on the George Washington Bridge.

I tune in Teterboro tower, and hear a comforting chatter. Lots of traffic still around. Off to my right, the sun has almost disappeared—not down below the horizon, but dropping swiftly into the haze. No matter; I have tuned in home base now. I thumb the microphone: "Teterboro tower, this is Tri-Pacer 9527-Delta, three miles north at 2,000." Like mother's voice, the answer comes: "Two-seven-Delta, report on your downwind leg."

There is the field, in lengthening shadow. I pass it to the westward, just as planned. In a long and sweeping turn, I cut my power back and circle around to get into the pattern. At 800 feet precisely I am downwind, and so inform the tower. They give me runway, wind, and altimeter setting. Ahead of me, a red Tri-Pacer starts to turn into its base leg. "Two-seven-Delta," the radio says, "your traffic is a Tri-Pacer turning base. Do you have him in sight?" "I do," say I. My friends are there, and they are watching.

The sun is down and out of sight as the end of the runway floats beneath me and my wheels touch down. The runway lights are already on; it is dusk down there. Ten minutes more, I think as I taxi off the runway, and I would have been landing in the dark. Student pilot Knauth, you cut that one very fine indeed.

There comes a time in every student pilot's life when this

will happen. What caused it specifically in my case, I do not know. It may have been the heat and a long day in the office, but I am not inclined to make excuses. Excuses never get a pilot home. I think what caused it is that I had nothing else in mind except to fly, just fly. And that is not enough—for flying is an ultimate thing, and whenever man approaches the ultimate, he must do so with respect, with knowledge, and with planning. Time, and the sun, and the speed of an airplane do not wait on whims; and I will never again fly on a whim alone.

7. *Flight on Paper*

Eight o'clock in the morning at Teterboro airport. I have come here to fly the Press Tri-Pacer for an hour, then take it to Morristown and pick up Jo Kotula, the artist, who will fly with me to the Piper plant at Lock Haven, Pa., for a day's work drawing brochures of airplane interiors. It is a nice trip coming up, but as I walk the long half-mile from the bus stop down to the hangar of the Safair Flying School, the prospects for it are getting gloomier by the minute. A light fog shrouds the airfield, a mist that is slowly turning to a drizzle. The dozens and dozens of airplanes that make Teterboro one of the busiest airports in the country stand in idle ranks, windshields and windows clouded by the moist, cool air, wings dripping, looking earthbound and miserable. Big planes, little planes, medium-sized planes, gay planes, somber planes, ex-transports converted to executive ships with big picture windows cut in their sides; ex-fighters and bombers gone incongruously civilian—Teterboro presents a mosaic of general aviation in the U.S.A. which even on this gray and moody morning cannot fail to be impressive.

At the Safair Flying School I check the weather. It isn't good anywhere except beyond Allentown, and it would be solid instrument flying all the way down there. I call Jo on the phone; no, he says, he doesn't want to fly instruments. Clay Collins, my instructor, sees little hope that it will clear today. And suddenly I have an inspiration: I am going to take my private pilot's written examination here and now—just walk right in and take it, cold.

Half a mile further down the row of hangars is an FAA field office where the exams can be taken any day of the week between nine and four-thirty. Maybe this is the way to do it—unexpectedly. I have sat for nights on end, working my slow way through the many sample examinations that can be had from both government and private sources, getting more nervous every time. It isn't the exam itself so much that scares me, it's the idea of taking an exam. Why, I haven't had a written examination of any kind for nearly twenty-five years! The more I've thought of it, the worse it's got. So now I will take myself completely unawares, and simply go.

Before that nervous panic can get me in the vitals, I ask Clay a few quick questions that pop into my mind. We work out a wind triangle problem together on the computer and he gives me some sound advice. "It's all right there in the books," he says. "There's no reason to get sweaty about it at all. Just make sure you *understand* each question before you try to answer it. They're tricky; they're deliberately tricky. And don't forget to calculate all your mileages and wind speeds and air speeds either in knots or miles per hour. They'll give you problems with both; you just get 'em all on one basis." Then he takes me down the line in his car, and while my resolve is still high and my stomach in its proper place, I climb the stairs to the FAA office and walk in.

And that is how, on a rainy day at Teterboro, I find myself flying an Aircadet 130, a totally imaginary airplane, on paper across half a thousand miles of Texas.

The FAA examinations could be anywhere, but for some reason or other the areas on the Cleveland Sectional Chart in Ohio and on the Dallas chart in Texas seem to be preferred. The principle of the examination is a simple and straightforward one: the candidate is faced with an imaginary cross-country flight in an airplane of a given type, capacity, and performance, and is asked to solve all the problems and answer all the questions attendant on such an enterprise. Everything comes into it—Civil Air Regulations, preflight inspection, preflight procedures, matters of weight and balance in loading the plane, take-off conditions, climb and flight problems, navigational problems, radio communications en route and on the ground, calculations of air speed at different altitudes and temperatures, calculations of ground speed in given conditions of wind and weather, weather problems, interpretations of weather forecasts, traffic control procedures, identification of airfields and their facilities, problems of fuel consumption, estimations of time en route and of points of no return, calculations of estimated times of arrival—the works. Considered as a single, all-inclusive examination, it is enough to chill the stoutest heart. Considered as a practical test of what a student has learned, it is one of the finest and most effective examinations ever devised in any field.

As Clay said, it's really all in the books, and the books are given to you. You do need to have read the books so that you will know where to look things up. And it is helpful if you have had some actual cross-country experience, for then this is just like flying.

So now I fill out my name and address on a piece of paper and am given my materials. I get a paperbound examination book, which presents me with my airplane, the aforementioned Aircadet 130, a four-place, high wing monoplane "of conventional design"; I get its performance figures—engine horsepower, fuel consumption, fuel capacity, gross weight, empty weight, take-off and landing characteristics at various altitudes and temperature, rates of climb ditto, stalling speeds flaps up or down and power on or off at various degrees of bank, and a graph showing load factors at various degrees of bank. I also get a regular Dallas area Sectional Chart with all of its vital information printed on the back (this is practically an extract of the pertinent Civil Air Regulations), a copy of the *Flight Information Manual,* the pilot's bible which contains vital information on everything from Air Defense Identification Zones to places where omni sets can be checked for accuracy on the ground or in the air, and a copy of the *Airman's Guide,* the bi-weekly publication which keeps the *Flight Information Manual* up to date. I am told where I must go—from Corsicana airport north to Sherman airport, thence via the Fort Worth omni to Graham, thence on a southwesterly leg down to Abilene Municipal airport, thence via Stephensville airport back to Corsicana, my theoretical home. I am given an answer sheet on which to block in my answers in the space provided—I get a choice of four, sometimes five answers—and a special soft pencil with which to mark the answers. The exam will be graded by machine, not by any human agent (again a rather chilling thought), so the filling out of the spaces must be carefully and precisely done. And now we are off and away.

First, the mundane matters of the journey. For purposes of this exam, I am a private pilot who weighs 190 pounds, a bit on the heavy side for me, and I am taking two business associates on this trip to discuss various matters of importance with clients. My associates weight 180 and 185 pounds, respectively, and between the three of us we have three briefcases, strapped together in the back seat, which weigh 48 pounds all told. That makes 16 pounds per briefcase, which is more than I would want to carry on the New York subway, but in the airplane I hardly notice them at all. In fact I find, when I total up all the weight, including 36 gallons of gas at 6 pounds per gallon and two gallons of oil at 7½ pounds each, that we could still carry our full load of 80 permissible pounds of baggage and be under our gross weight limit. This gives me pause, briefly, until I recall that my Aircadet is built to carry four people and, after all, we are only three.

That's one of the early questions, but already I am absorbed. Around me half a dozen other sweating candidates are equally engrossed, some in a private written examination like me, some in a commercial or instructor's written, a couple of godlike souls in the toughest one of all, the examination for an air transport pilot's rating. Two FAA clerks chatter about matters of office routine; outside in the rain an executive DC-4 warms up with shattering roars, taxis out, and takes off into the low-hanging clouds. I light another cigarette and bend to my task.

It is a long time before I get into the air. I have to go through my airplane from spinner to tail before I can even climb in and start up the engine. I have to know what documents must always be on board. I have to know when it will come up for its next periodic inspection—it was last inspected five months ago and has since had 110 hours on the engine. I have to say what things I would check on the engine beyond the usual routine of checking the oil, draining gas from the sediment bulbs to see that it is clean and pure, and shaking the propeller to see that it is on good and tight. This is a matter of judgment, not rules, and I ponder it for a while. But fortunately I know that an airplane engine doesn't have an automatic choke, and that eliminates two of the possibilities. I finally settle for what seems only sensible:

checking the engine mount for security, checking the spark plug leads and terminals for cleanness and tightness, and inspecting the exhaust system for tightness and cracks.

I also have to figure out whether, at an assumed airport altitude of 2,000 feet and with a 2,000-foot runway at my disposal, I can take off and clear a 50-foot obstacle when the temperature of the outside air is 70 degrees. I have to interpolate this from the airplane flight manual charts which give me my take-off and landing distances at altitudes from sea level to 6,000 feet and temperatures from zero to 100 degrees Fahrenheit at 20-degree intervals. It seems to me that I can make it, because although my take-off distance to clear said obstacle is something over 2,000 feet, it is also indicated that my ground run in each case is only about 40 per cent of the total distance, which leaves me room and to spare on the runway. I hope I'm right.

How much higher is my stalling speed in a 60-degree bank than when I am flying straight and level? This is a very vital question, because a student, or even some fairly experienced pilots, are prone to forget that an airplane in a bank stalls at a much higher speed. In the case of my Aircadet, for example, the difference is 24 miles per hour—59 miles per hour straight and level, a good, fast 83 miles per hour in such a really steep turn. I am extremely glad to be reminded of this; hours of flying in the pattern and even later on cross-country trips have accustomed me to the idea that 83 miles per hour is a nice safe speed.

With all this behind me now, I still have to tackle the weather. And here the pilot—student, private, or otherwise—really joins hands with the pros. From Piper Cub to Boeing

707, everybody gets the same type of weather reports, becomes a privileged customer, free for nothing, of the tremendous meteorological network of the FAA which spans the country and reaches out twenty-four hours or more into the future to let the flier know when he can safely fly. I am still awed by the fact that anywhere, anytime, I can pick up the telephone and call any one of more than six hundred FAA weather stations and ask them to let me know what the weather is and is going to be at any given place or time, and they will tell me. I can also get more detailed information in area forecasts, put out four times daily and covering a twelve-hour period each; in terminal forecasts which give me information on ceilings, cloud heights, cloud amounts, visibility, weather conditions, and surface winds at various airports; in hourly sequence reports and forecasts of winds aloft, all of which go out across the country on teletype circuits. Furthermore, once their relatively simple abbreviated codes have been mastered, the story they tell is a very clear one. But still, weather being as important as it is, this whole part of my preflight routine really brings the perspiration to my brow.

For one thing, a cold front is passing through the area where I will shortly be taking off and through which I will be flying. A detailed weather map gives me a precise picture of just what this front is doing to the weather in these parts, but I am not entirely sure I understand its story. This is one question where I erase three times—a risky procedure in an exam which is graded by an electronic brain—before I finally plump for caution. Between taking off at 7 A.M. "with plenty of margin for safety" and taking off sometime before noon when I can "be sure that weather conditions will permit safe flight," I decide in favor of taking off sometime before noon. I am not quite certain whether that front was actually past Corsicana or whether it was just about there at the time the weather map was drawn. These maps are not specific as to locations, and I would rather be safe than sorry. Whether I was right or not I'll probably never know—the machine may grade the answers well enough, but it offers no subsequent explanations.

It also seems to me, after careful study of the sequence reports for areas that I will be flying through (these are issued hourly or, in rapidly changing conditions, even more frequently), that I will have generally improving ceilings and visiblity as I travel north from Corsicana to Sherman. One of the alternatives given me is to cancel my whole flight here and now, but since I have some twenty-five questions still to go, I feel this would be hasty. And so, at 0925 hours under a gray but lightening sky, with a 1,000-foot ceiling and 15 miles visibility, I finally take off, clear that 50-foot obstacle and climb.

From Corsicana to Sherman is something over a hundred miles. One of the first things I have to do is look for check points on the ground so that I can keep definite track of my progress. Here comes one of the very few questions which I do not consider entirely fair. I am given four different lists of check points to choose from—at least three of them make pretty good sense. There are quite a few towns, highways, railways, towers, lakes, and other landmarks on this route which can serve my purpose; it seems to me that it is entirely a matter of individual judgment which ones might be chosen.

I mull over this list for quite a while before I decide on the one which has its points most evenly spaced, and also includes the most prominent ones, among them a large-sized reservoir which I will be directly overflying.

I also make my first acquaintance on this flight with an omnirange station. Dallas—DAL—is there and I am asked how I will identify it. Its frequency is given, but this is not enough; frequencies are subject to change and must be constantly checked in the *Airman's Guide.* The code signals given do not fit it. The possiblility is propounded that it may have voice identification—I can hear in my ears those lugubrious tones that seem to be characteristic of this type of range station: "Dallas OMNI!" Repeated endlessly and interminably, it seems like a funereal dirge by comparison with the cheerful chirping of the code signals. I check the *Airman's Guide* for this and, lo and behold, there it is: Dallas—M-*BVOR, with that asterisk signaling that it has an automatic voice. With the dirge still sounding in my ears, but happy now, I fly on.

But I am not long left to fly in peace. The check points now come into their own: I am told that I passed so-and-so at such-and-such a time, and am now, ten minutes later, coming up on this-and-that. What time can I expect to be in Sherman? I get my plotter and measure off the distance between the two, being extremely careful to pick the side that is calibrated to the sectional scale. In ten minutes, I find, I have gone 17 miles, which figures out to be an average speed over the ground of 102 miles per hour. Not bad, considering I am bucking headwinds. It is easy now to estimate my ETA in Sherman—so-and-so many miles still to go at this same speed, and there I will be. That is, if the winds don't change.

But on this leg they don't, and we arrive in Sherman in time to transact our morning's business and go on. Should we refuel here? Well, that question was decided back in Corsicana, before we left the ground. No, we should not refuel here, because we have been told that we can refuel only once—and if we fill the tanks in Sherman, we will not have enough to complete the rest of the trip and still have our required reserve of one hour's flying. We burn nine gallons an hour; we carry 36 gallons of gas, and if we want to keep an hour in the tanks for emergencies we have a maximum range of 366 miles at an average ground speed of 122 miles per hour. Considering that on this leg at least we are 20 miles per hour slower, the reserve is a smart precaution.

From Sherman to Graham we get a little omni problem. We are going via the Fort Worth range, so I am asked what radial I will be flying on and whether the indicator should read TO or FROM. This is a question which can trip the unwary —I remember, however, that I plot my course through the center of the range and out the other side, and take the westward bearing indicated there, and not the one on the side nearest to me. And despite the fact that this bearing is away from the center, seen from my side, the indicator still reads TO. Logic backs up my recollection: I am flying west, my heading is around 240 degrees, so that must be the one. And when I cross the range station, my indicator will waver and swing over to FROM.

The tripper-upper here is that a hasty look will indicate one should be flying on the heading nearest to the plane, which would be 60 degrees TO. But in that case, we would be flying east, in the opposite direction, and we would never get to Abilene, not in a million years.

Flying toward the Fort Worth omni, I am also asked to consider the wind at my altitude. It is blowing 18 knots from 15 degrees, from behind and to my right. At an indicated air speed of 120 miles per hour and an outside air temperature of 14 degrees Fahrenheit, where does this leave me? Must I correct to the right (or left) to compensate for drift and keep the needle on my Omnigator centered? What is my ground speed with this wind?

This one is full of tricks. To take them in order: No, I need not correct to keep the needle centered; as long as I keep it there, staying on this radial, I am automatically compensating for wind drift. As for my ground speed—well, first I must figure my true air speed at my present altitude of 4,500 feet and a temperature of 14 degrees Fahrenheit. Slow and easy is the word here; it is all neatly worked out on my computer, but the tricks are carefully built in. Since the computer works in Centigrade, I have to convert the temperature from Fahrenheit first. That done, I calculate true air speed. Then comes the wind—I have that all figured out before I suddenly remember that winds aloft are always given in knots, so I backtrack and work that out in miles per hour. Then I can figure ground speed, wind drift, and, just for good measure, wind correction angle, all on the computer. And when they ask me what time I estimate for Graham, I can tell them: in time for lunch.

Well fed and rested, we take off from Graham, then, and head for Abilene. On this leg I am kept busy with problems of figuring compass headings. This takes into account true heading, figured from true north with the plotter; magnetic variation, which here in Texas must be subtracted rather than added as I do back east (the variation here between true north and the North Magnetic Pole is an easterly one, and "east is least and west is best" is the well-worn rhyme that spurs my recollection); then I have to figure in the deviation in my magnetic compass, which is supplied with other pertinent data on the airplane, and finally, once again, the wind correction angle. I also note that Abilene is in a control zone—a thousand-foot ceiling and three miles visibility are the minimums here—which brings up questions of landing procedure. Graham, for example, is in a control *area*, where rules and regulations apply only from 700 feet on up. But one way or another, we finally make Abilene.

And now, as the afternoon is wearing thin, we leave picturesque Abilene to bask in the sunlight filtering through its 5,000-foot broken ceiling and head for home. Abilene omni lies directly behind but somewhat north of us; it is postulated that we prefer to fly to Stephensville on an outbound leg of that. I am given an omni radial, and then a check point along the leg I marked out previously; if I fly this radial, where will I be in relation to that check point? It works out that I will be somewhat north of my original track. Okay, allowing three minutes for climb to altitude from Abilene, and so-and-so-many minutes elapsed time to that check point, when will I be overflying Stephensville? I have to figure out the mileage on my new course now, and the answer leaves me right smack in the middle between two alternative answers. So I figure and refigure, finally concluding that I have been a little stingy with my miles—they have placed me "due north" of one town and "at the northern edge" of the other. Being extremely careful, and recalculating with exactitude, I finally choose my answer. Once again, I hope it's right—this is the only other question that I cavil at, because it seems to me it could be either way. Whichever it is, Stephensville passes below me; I can practically smell dinner now at Corsicana.

There doesn't seem much else that a pilot could be asked to figure after such a flight, but there is. The sun sets at Corsicana at 1920 hours. With that 15-degree, 18-knot wind still dogging me, will I make it before dark? Only this time the wind isn't 18 knots anymore; it's 20 miles per hour, just to see if I have wit enough left to spot the difference between knots and miles. But there can only be one possible answer to this one—I'll make it with 14 minutes to spare. The other alternatives are that I will make it three minutes before sunset, three minutes after, or (I believe I recall) by the light of the Texas moon, and these are all too far off to be considered.

And so we land at Corsicana as the sun sinks in the west, a tired but contented trio of businessmen, one of whom, the pilot, has lost at least ten pounds. One more obvious question remains: What must I absolutely do now, if I do nothing else? Well, close my flight plan, naturally. Otherwise within an hour Search and Rescue will be looking for me, thinking I have crashed. (But this obvious question is the one they don't ask, and I have forgotten now what they did. In any event, we're home).

The room is blue with cigarette smoke as I raise my head and rub my weary eyes. I hand in my finished examination and stumble down the stairs, through the cavernous hangar with its assorted mixture of huge and little airplanes, and into the cafeteria. I have been flying—excuse me, writing and figuring—for six mortal hours. A sandwich and coffee is in order, and make it a big one, please.

Back at Safair, Clay Collins is leaning against the hangar door, studying the gray and moisture-laden sky. He greets me with commiseration; he is looking forward to taking his air transport pilot's written soon. I breathe the cool air deeply. "Hey," I say suddenly, glancing heavenward, "couldn't we fly? It looks like we've got the minimums!"

Clay shakes his head. "No," he says, "We've only got eight hundred feet. And you know what we've got to have."

"Yes," I say, "a thousand feet and three miles in a control area."

"Control area?" says Clay. "Now you know better than that. We're in a control zone here, and you know what the difference is, don't you?"

"Of course, of course," I answer hastily. "Good Lord, yes." And I walk sheepishly off to the bus and home, leaving the very real Tri-Pacer sitting under an 800-foot ceiling and a visibility of quite a lot less than three miles.

8. *The Ultimate Flight*

Somewhere in Europe, as I write these lines, a little blue-
and-white airplane is winging through the skies on journeys unknown to me—an airplane which in a brief span of time became an irreplaceable part of me. She flies under the letters

D-GARY, a German registration, but she is American-made—a Piper Apache, with room for five in her small blue and silver-trimmed cabin and two 160-horsepower Lycoming engines to speed her on her way. She will have a lot of hours on those engines now, logged in flights that have taken her all over two continents (a Pan-American pilot friend of mine not long ago reported seeing her on the ramp at Lisbon, and later I had word from someone else that she was touring Africa); but when I first knew her, she was gleaming, factory-new. To me she was only an abstract entity then, a ship to be delivered to a customer in Düesseldorf, West Germany, by the simplest, cheapest method—under her own power, on her own wings. What set her apart for me is that I went on that flight, and in the course of it, I came to know D-GARY as a man knows his best-beloved, his most trusted friend; and with her, her pilot, Max Conrad, probably the greatest flier of our age.

I had heard a great deal about Max Conrad. Other fliers spoke of him almost in terms of awe, as in earlier days they might have spoken of Charles Lindbergh. He had flown the Atlantic alone in small aircraft more times than any other man in history; he was nearing his fiftieth crossing when I met him. His first Atlantic flight had been made in 1950 in a Piper Pacer, forerunner of the Tri-Pacers which I flew, but with only 125 horsepower as against the 160 horsepower in the current models. In that tiny ship he flew the ocean twice! Two round trips, via Labrador, Greenland, and Iceland; hours and hours alone over the empty sea, with just one engine and his own indomitable faith in that engine, himself, and God. And after those pioneering journeys came almost half a hundred more in various planes; in Apaches and Comanches, a Beechcraft once, and another time a converted P-38 Lockheed fighter plane—jobs to be done in his career as a one-man ferry company specializing in long-distance deliveries. These many, many flights, including one across the Pacific to Formosa; these and his phenomenal flying record of 36,000 hours in the air had given Max Conrad the aura of a living legend, a flier's flier, a type that passed from aviation's increasingly commercialized and matter-of-fact world with the end of the Lindbergh era.

Now I was to fly the ocean with him. We had met and talked only a few days before, at luncheon with Bill Strohmeier, and Max had proposed the trip as casually as if it were an invitation to run up to Martha's Vineyard for the afternoon. "I've got an empty seat in the airplane," he had said. "I get kind of lonesome on these trips. Why don't you come along? We'll leave on Friday; you could be back by Monday." Who could have refused?

I liked and trusted Max Conrad from the moment I first met him. A quiet, graying man in his fifties with a manner so diffident as to be almost self-effacing, he nonetheless had a quality about him that inspired complete confidence; a combination of faith, idealism, and awareness of himself such as I have seldom encountered. If one quality exceeded all the others, it was that he knew his limitations. I have since come to realize that this is often mistaken in him for modesty; actually, it is perfectionism. Every man has his limitations, but not many recognize them as clearly as Max Conrad. What he has done, however, in thirty years of flying is to push his limits, by dogged persistence and an extraordinary self-discipline, far beyond those of most men. Where others might relax and rest on their achievements, Max never did, and probably never will. Like a climber who sees always one higher peak looming ahead of him, Max always looks back on a long flight with an eye not on how well he did it, but on how he might have done it better. This was, eventually, to lead him to the two longest solo flights ever made by any man; but these were still in the future as we discussed our trans-Atlantic journey that day at lunch in the Wing's Club in New York. And I, impressed as I was with this quiet, gray-haired man who spoke so casually of flying a Piper Apache to Germany, had yet to learn his true worth in the flight which is to me the ultimate flight, the greatest adventure of my life.

Here is the log I kept of that journey across the sea.

ACROSS THE ATLANTIC WITH MAX CONRAD

FIRST LEG, Flushing–Norwood, Mass.

Seven o'clock on a rainy cold morning. We are driving out to the little airport in Flushing where the Apache is parked. We check the ceiling as we go, measuring it against the skyscrapers. Not good—RCA Building's top is in the clouds, ditto the Chrysler Building, its needle tip disappearing into dirty gray scud. Yet we are anxious to be off, lest conditions get even worse. As we cross the Triborough Bridge the sky seems to lighten somewhat. Max figures we have a bare minimum—a thousand feet—but certainly no more.

It is drizzling slightly when we reach the airplane. The airport is silent and deserted. No time for farewell, just a quick kiss for my wife; then I'm in, putting on my seat belt, shutting the door. Engines start, first one and then the other, and before the second one is even fully started we are already swinging around, heading up the runway. Rain streaks the windshield; Max peers right, left, all around, constantly searching from side to side. We turn onto the runway, he guns the engines. "Well, I guess we're on our way," he says, and off we go.

We don't climb out at all. The big Consolidated Edison gas tank is above us as we go swinging past it, turning left, out over the island. Visibility is good enough here, but the ceiling is right down close, that's for sure. We are at 900 feet and the clouds rip past just above us. We reach for 1,000, and an occasional wisp whips past the windows. We hare out across the suburbs just under the ceiling. Houses, stores, highways zip past below, then water appears on the left. At Great Neck (I judge it's Great Neck) we swing left over the water. "We'll go over the water," Max says. "No towers there, so we can stay low." So now we are paralleling the shore line which I can make out clearly, if grayly, on my right.

This man, when he is in the air, is in the familiar place. Once off the ground, he seems to relax, to be at home. He flies with absolutely effortless ease, without any apparent conscious application; he flies the way other men walk; he seems literally to have been born to it. I turn around and watch him for a moment. He sits hunched forward, one hand on the wheel, feet gathered beneath him, his eyes constantly searching ahead and to both sides, every sense, his whole body seeming to reach out into that gray and turbulent world through which we dash with the urgency of a great bird seeking shelter from an approaching storm. Max doesn't seem to be in the airplane at all; he seems to be out there, sniffing the air, probing it, trying to sense what's ahead; and what his hands do on the wheel is mere instinctive reaction to what his senses feel outside. I have never seen anything like this at all, but now I know what manner of man gave birth to that trite phrase: "He was born to fly."

We are doing 155 mph indicated, 23 inches on the manifold pressure. We are full on both tanks; the engines are humming sweetly; we have plenty of range and plenty of time. I sit back and relax, keeping an eye on my side and the shore, trying to make out where we are.

Oyster Bay goes by in gray scud. Max turns the dial of the VHF; we pick up an omni beacon. It has a most familiar rhythm, one that I have heard often before: .... ..—. —.. Nantucket? Providence? Look it up—it is HFD, Hartford. We are heading due east; the beacon is ahead to our left. It is 7:40—we have been flying fifteen minutes by my watch (providing I got the time right in the hurry of the take-off).

Now we probe upward. We reach 1,000 feet and the ceiling is still above us; 1,200 and we are still okay. Underneath is only water; we have left the Long Island shore and are angling across. Now clouds whip past below us. Briefly we are right in the soup. In and out again; on the left I make out a bay, a breakwater, houses. New Haven? The clouds close in again.

We fly on for what seems a long time. Pieces of shore line whip into and out of my vision like pictures driven by the wind, fragments, ragged and incomplete. I begin to wonder if that rocky bit below with the wide estuary beyond could possibly be Narragansett Bay. It couldn't, not possibly; we haven't been in the air for half an hour yet. Then suddenly we are completely in the clear beneath the dark gray sky. A water tower, a railroad station, a highway, a two-lane throughway—and a building marked Clinton. Now I know where we are precisely. We head up the turnpike and I can just make out the exit to Horse Hill, and the road; but before I see the house where my sons live the clouds close in again. It is 7:50, and we have passed Westbrook, Connecticut, at 1,400 feet.

A voice comes on the radio, clear and loud, with the weather report. Providence has 3,500 feet, 1½ miles; Westover 6,500, broken, 5 miles; La Guardia 1,000 feet, 1 mile. Well, we got away none too soon, and it looks good ahead.

Eight o'clock. We are at 1,200 feet, rain drumming the windshield, clouds above and below with only an occasional break. The omni beeps steadily in our little, silent world. Max puts the carburetor heat on every now and then, just to make sure we don't ice up inside. Then, abruptly, clouds are all around us; we fly in a blank, gray void.

On instruments, we start climbing. Seventeen hundred feet, and the sky seems to be vaguely lightening ahead. At 8:15 we are at 1,800 and run into rough air. "Good sign," says Max. "It should be clearing pretty soon now." Then it starts to rain again.

Boston comes in with the half-hourly weather report. Boston has 1,800 broken, 5,000 overcast, 7 miles. Portland has 3,000. Providence has 800 estimated ceiling, broken, 3 miles visibility. In half an hour it has closed in from 3,000 feet to 800. "This stuff is moving pretty fast," says Max.

The clock says 8:23; I look down and suddenly I see a wisp of ground. I point it out to Max, and he nods. Then suddenly a whole landscape, snow-covered. He banks steeply, first left, then right, coming down in a fast spiral. At 700 feet we pick up the turnpike again. We fly right over it, bumping in the wind, turning where it turns while I try to make out road signs. "Providence must be somewhere real close now," says Max. I am sure it is, because the turnpike ends in a welter of blinking lights and now we are flying up the snow-covered emptiness of a turnpike-to-be.

Hills loom ahead. The ceiling gives way grudingly; we are at 1,200 again and still in the clear. There is a large hill on our left, topped with a tower. "Blue Hill," says Max. "The airport we want is just over there." We turn and head toward where he pointed, and sure enough, there it is—a large white runway. A gentle turn right, a straightening turn left, and we come in. Wheels come down, snow flashes past and we are on the ground so gently I hardly feel it. It is 8:35; Norwood, here we are.

Max is apologetic. "I shouldn't have let myself be caught like that," he says. "But it just goes to show you how you can get caught. What happened was that we were between two layers of clouds and the two came together. That generally means that your ceiling underneath you is lower too, because the one cloud layer forces the other down. You know ninety per cent of the people who do get caught get caught just that way. I knew it when we got that weather report from Providence—down to 800 feet in half an hour."

I ask him what he would have done if we hadn't come into

the clear at Norwood. "Well," he says, "it was going through my mind. I decided if we didn't, I'd turn and head out over the ocean until I got on the Boston low-frequency range. Then I would just come in on the beam, letting down over the water. The only thing you can do if you get caught—and you really never should—is to home in on some big airport and if necessary, make an instrument approach or have them talk you down."

D-GARY is in the hangar now, her nose off, her cabin stripped while the mechanics work on the autopilot and the glide slope radio we brought along. The autopilot is not working properly; Max had said this morning that the pitch control (the fore-and-aft, up-and-down control) wasn't working at all. On the way up here the lateral control (the side-to-side steering) went out too. It is all a new installation, and it will be fixed before we leave; not that he needs it for the flight particularly, but so that it will be right when the plane is delivered. Meanwhile he is at work installing the two big tanks in the cabin. They are two large, square metal boxes, each holding 80 gallons, connected with a cross-feed which in turn is connected with the cross-feed of the regular tanks. The way it looks now, tanks and radio will be in by tomorrow afternoon; then we will give everything a thorough check flight and be ready to go.

Monday dawns bright, clear, and cold. Saturday's storm moved northeastward Sunday morning, leaving the Boston area clear yesterday afternoon; by this time, Max surmises, it will be well out of our way and conditions right now look ideal for the crossing. We are up at six, out at the airport by seven after a good breakfast; now we pack our things into the airplane, pull D-GARY out into the frosty morning air, and are ready to take off for Logan Field in Boston. There we will fill all tanks, get the latest weather reports, file our flight plan, and check through Customs before we are really on our way.

Quick good-bys, door closed and locked, engines started, and we taxi down to the end of the runway. We line up there and check out both engines, then move into position for the take-off. "You might as well start working now," says Max, and motions to me to take the throttles and the wheel.

I am slightly petrified by this. Holding an Apache on course when you are three or four thousand feet in the air is one thing, taking it off is quite another. But now I get a demonstration of the Conrad no-talk instruction routine. He folds my fingers over the two throttle handles and motions me to push them slowly forward. I do so, and the Apache starts to roll. At two-thirds throttle he motions "that's enough"; by that time we are speeding smoothly down the runway, and I can feel the lift begin to take hold of her wings. Now a slight, gentle beckoning motion of the Conrad hands—lift her off, feel her off the ground. I pull back slowly, and up she comes. "Gear up," says Max, and I fumble for the landing gear handle, holding the ship straight and level meanwhile. I can't get the catch released; my fingers work around on it; finally, up it comes. We climb out, slowly, turn left at about a thousand feet and head for Logan. Max gets landing instructions from the tower—Runway 22—and down we come in a wide spiral, losing altitude fast; full flaps as we cross the fence; she balloons, slows, sinks and we are down, fourteen minutes out of Norwood.

There is lots to do at Logan. First we fill out papers for the Customs, and Max presents his various documents—export license, ferry permit, and so on. Next a quick stop for Max at the chapel of Our Lady of the Airways; then we go over to the control tower building for Operations and Weather. There is a full briefing ready for Max at Meteorology—a Forecast Folder with weather chart and forecasts all the way to the Azores. (The chart will be our navigational chart for the trip over—that and no more.) Air Traffic Control gives us our flight instructions—we are cleared to the EEL Intersection, a point in the sky about 120 miles out where Boston Control ends and New York Oceanic Control takes over. We file our flight plan in another office—Boston to Santa Maria, Azores, VFR at 7,000 feet, Estimated Time of Arrival 0540 Greenwich time—something over fifteen hours from now. Then back to the Customs again. Max wants five minutes undisturbed to go over the airplane once more—"There will be no place to land between here and Santa Maria, so I want to have one more look at everything" —and I wait, chatting with the Customs officials, until I see him wave. Then I shake hands, hitch my cameras onto my shoulders and, with a strange sense of elation, anticipation, and fatalistic acceptance of what is now my inevitable lot, walk out to the airplane and climb in.

We now have 152 gallons of gas in the two big tanks in the cabin. We have 36 gallons more in the auxiliaries, 72 more in the mains. We have two one-man rafts, two Mae Wests (they are lying at our feet, ready for quick use), a paper bag of sandwiches and cold chicken, fruit, candy, and two thermoses of coffee, one regular, one black. We have our luggage—Max's briefcase, my overnight bag, my camera bag, my heavy jacket. The back seats are piled on top of the cabin tanks—with our blue and silver interior trim and all this stuff, we look incongruously like a suburban station wagon inside.

We are heavily loaded, but we have fuel enough to go all the way across if necessary. The air is bright and calm, two airliners are poised on the runway ahead of us as we taxi out. D-GARY looks small but capable, and she feels good. We run up our engines one last time, get take-off clearance from the tower, Max motions for me to take over—"Full throttle this time, all the way"—and in what seems like an immense silence I push the throttles forward and we start to roll.

The take-off is surprisingly easy. I watch the air speed, but when it gets to 80 I switch my attention to the runway. The Apache doesn't even seem to feel this load. Max beckons, ever so gently; I ease the wheel back. She is a bit reluctant; I ease it back some more. A few short bumps and we are air-borne. I hold her level a foot or two off the runway and reach quickly for the gear handle. The lights blink on, one, two, three; I am easing her up into the sky. It is 1424 Greenwich Time, which is our time from here on in; 0924 Eastern Standard Time, just an hour and a half after we arrived at Logan.

Climbing slowly, I try to hold her at 120 mph. For the first time, I really begin to sense how delicately an airplane can be handled—*should* be handled. The least little tug or relaxation on the wheel is immediately reflected in the air-speed needle. Max motions me to start a gentle turn to the left, to get on our heading; at once the air speed creeps down toward 115, 110, until I have found the right amount to gently push the stick forward and bring it up again. Slow and gradual as this climb is, it has the added feeling of peculiarity of climbing with the wheel held forward instead of eased back: we have such a load of gas in the cabin behind us that Max trimmed the stabilizer to full nose-down position before we took off, and even at that the tendency of the airplane is to climb.

In half an hour we have reached our cruising altitude of 7,000 feet. We are now far above a field of scattered clouds; ahead of us, they thicken into a solid floor. The ocean is occasionally visible below, a distant, steely blue. Off to the right, a contrail arcs through the sky, fading to fuzziness at its landward end. Otherwise there is no sign of any living thing, and for the first time I feel a little stab of sudden realization of where we are and what we are doing, and how utterly alone we are here in this little plane above the empty sea.

At 1500 hours, on our heading of 115 degrees, we check our gas system. Max shuts the feed off the right engine; it immediately goes out, the airplane rocks downward to the right, I plug the electric fuel pump lead into the cigarette lighter, and the engine starts right up again. We check in the same fashion on the left engine; okay there too. From now on in, via the cigarette lighter plug and the little electric fuel pump, we are on the cabin tanks, which should last us to the Azores. With the cabin tanks leading into the cross-feed of the regular tanks, all the other tanks are shut off and the gauges register zero; when the cabin tanks are empty, we simply unplug the cigarette lighter, shut the feed valve from the cabin tanks, turn on the regular tanks, and proceed as normal.

We are now holding course with the autopilot. Since this works in conjunction with the directional gyro, Max has turned the gyro to zero and then switched on the autopilot. From now on in, it will hold the airplane on that zero heading as long as the autopilot is on—for instance, to change five degrees to the right, Max cages the gyro, turns it five degrees to the right, uncages it, and the autopilot promptly puts the plane into a turn until it has found the zero heading again. In a way, this simplifies matters—even with the autopilot off, it is easier to hold a plain zero heading than to hold one of, say, 117 degrees. At least, I find it so, and will remember this. The main thing is, of course, that the gyro must be set at zero only when the plane is firmly on its proper course by the magnetic compass. Steer 117 magnetic, wait till the compass has steadied down, cage the gyro and turn it to zero, uncage it and there you are.

We keep the ADF, the automatic direction finder or radio compass, on a zero heading too; and now we tune in Yarmouth, off to our left on the lower tip of Nova Scotia. It comes in loud and clear, a musical signal note, and the ADF needle swings to 310 degrees, pointing to where Yarmouth lies off our port bow. When the needle is at 270 degrees, at right angles to our course, we will know we are directly abeam of the Yarmouth range. Thus we can check our forward progress, station by station, all the way up the coast of

Nova Scotia and Newfoundland while we proceed diagonally away from the coast line out into the wide Atlantic.

This is our weather situation for the crossing: there is a big low-pressure area centering somewhere around Greenland; on the map it is drawn like a great irregular pool with concentric waves reaching down toward us. Around this winds are circulating counterclockwise. For us right now these winds are northwest winds, blowing from behind and from our left, pushing us partly southward, partly ahead. As we proceed they should change to west winds which will be tail winds for us, then gradually to southwest winds which should push us north again. Thus, as in the case of the magnetic variations, the winds should tend over the entire crossing to cancel each other out, and this sort of figuring forms an important part of Max's navigation. There's a high to the south; we ride between both.

1540 hours. We call Boston—twice—but fail to raise them. So we have now lost Boston Control; we are well beyond their range. We try New York on the Sunair high-frequency radio, Max's own, which he has wedged between the seats. He lets out the trailing antenna, puts on his headset. Okay, here is New York, loud and clear.

Our floor of cloud is breaking up now; ahead and to the left there seems to be clear ocean. Off to the right, far distant, a pale tan bank on the horizon, is the weather of the high-pressure area to the southward. To the left, abeam and slightly behind but much closer, is the trailing whiteness, veiled and ragged-edged, of yesterday's storm going up toward Newfoundland. Gander would have been very difficult today. "Tomorrow," says Max, "it would be wonderful for a flight to Shannon; tail winds all the way."

1552 hours—we start to pass through the Yarmouth beam. The needle has slowly swung to 270 degrees, pointing straight left at right angles to our course. The signals beeping musically through the loudspeaker blend gradually into one continuous hum—A's and N's canceling each other out to indicate the center of the beam. The clouds below are widely scattered now, the sun shines bright and warm into the cockpit, the air outside feels cold and pure and we feel warm and snug. Engines are running beautifully, throttled down, quiet. I eat a peanut butter and jam sandwich and we both have a swallow of coffee. "I don't know when I've had it better at the start of a crossing," says Max.

I take over again for a while and Max checks weather reports from Canada. At 1605 we are solid on the Yarmouth beam, so we switch over to the next one up the line at Dartmouth, halfway up the coast of Nova Scotia near Halifax. The signal comes in loud and clear.

Here is our instrument situation; we have cut off the manifold pressure lines, oil pressure lines, and the priming lines, eliminating these as possible sources of trouble. We have our ADF, dual VHF radio with omni navigation facilities, instrument landing system including runway localizer and glide slope indicator, and Max's Sunair HF radio for long-distance transmitting and receiving. We have magnetic compass, gyrocompass, artificial horizon, electric turn and bank indicator, rate of climb indicator, sensitive altimeter, suction gauge, and the rest of the standard instruments.

Power settings: throttle, props, and mixture are all in a

line about halfway down the quadrant. The mixture has been leaned down quite considerably. Max figures we are running at about 2,200 r.p.m. which is around 20 inches of manifold pressure. The ship now trims almost level with the nose trim still at full DOWN position; we are slowly using up some of that tail weight in the cabin tanks.

1629 hours. We have now been flying a little over an hour. We try Sable Island on the ADF and raise them at 330 degrees. They transmit one minute after the hour and half-hour in clear weather, so at first we wonder if our clock is wrong; but no, at 1631 we get them again. "Must be foggy up there," says Max. "Seems very likely it should be." We down here are CAVU (Ceiling and Visibility Unlimited), with blue sky above, blue sea beneath, occasional scattered clouds and a pale daytime moon hanging transparently in the sky ahead. Our air speed is 130 mph, our altitude 6,800 feet, and everything is just fine. The outside temperature is slowly creeping up from 20 degrees Fahrenheit toward the freezing mark.

1650 hours, and we are coming up on two thin bands of clouds far below us. We have just reset the gyrocompass again; it needs to be reset about every half-hour.

1655 hours. We see a ship off the port bow, playing hide-and-seek with the little fluffy clouds that are now beginning to appear. At first I think it is two ships sailing together, a little one and a big one. Then it looks like two planes, flying low; but what would two planes be doing way out here at wavetop height? They duck in and out of the clouds, seeming to change their shape, but now it comes clear—it's a liner, and quite a big one, trailing a white wake.

I ask Max whether we could raise the ship on the radio. "No," he says. "For some reason the sea and the air don't mix as far as communications are concerned."

1701 and here is Sable Island, right on time. We tune back to Dartmouth, and find we are passing through the beam. Projecting the Dartmouth beam down to our course, it looks as though we were a little ahead of our estimated position. Max calls New York to give our new position, checks back to Sable Island on the ADF, and a rough fix places us indeed quite a bit ahead.

I am flying again, and I find I now have to hold the wheel slightly back to keep straight and level, so I trim the nose up a bit. It strikes me that we will be doing this from now on until the cabin tanks are empty—bit by bit that red dot on the trim setting will slide back along the trim position line, like an unofficial gas gauge registering our fuel consumption, until in the Azores it will be in normal trim position.

The clouds are slowly beginning to form up again below. They are still very broken, but they are thickening and forming a tufted, ragged army in scattered ranks to the horizon far ahead.

1816 hours. We call New York—three times—no reply. On the third try Max gets an answering call from "McKinley," loud and clear. We have no idea who McKinley is, but he relays our new position to New York for us. Max has again revised it forward from earlier estimates; we are doing well.

CAVU with slowly thickening cloud floor. The sea seems to be roughening up; there are whitecaps on the dark blue water. We are still flying comfortably right between the two weather systems, the high to the right, the low to the left, but drawing away from the low. The sun is moving behind us now.

Power settings: 2,000 r.p.m., manifold pressure 20 inches (estimated), mixture leaned still a bit more, air speed 130 mph, altitude 7,000 feet. Max figures we are running at just about 50 per cent power.

For the last half-hour or so I have been slowly milking D-GARY up to her cruising altitude from about 6,000 feet. This is a wonderfully delicate job, as I have been trying to do it without either Max or the engines noticing the climb. I watch the air speed dial, the altimeter dial, the rate of climb dial alternately, watch the horizon and very, very gently ease back the wheel with my fingers. Inch by inch we go up; then I relax too much and we slide down again. It is like climbing a slippery hill. I trim the nose up just a touch, less than a quarter turn, and finally we make it, with a real sense of triumph on my part. Max meanwhile has been checking here and there on the radio. At 1940 we are to call New York again.

The cloud floor ahead is slowly slanting up toward our altitude. The sky is still clear blue above, but a deeper blue now, and the sun's rays slant from behind.

At 1848 hours we get our first sign of the weather ship, *Ocean Station Delta.* It is patrolling the seas near 40 degrees west and 45 degrees north. We see it as a brief flicker on the ADF dial, which has been tuned to its frequency for some time now; almost as soon as we see it, it is gone again. It is odd, in this world of radio communication, how one comes to feel with those invisible radio waves, sense with the instruments—it is as though I actually had caught a brief glimpse of this unknown ship somewhere far below, plunging in the dark seas, sending out its pulsing beacon signal. Delta transmits regularly at H-plus-5 to 10; 20 to 25; 35 to 40; 50 to 55; so we wait now for the 50-55 signal. She is pretty far away, on our left between Newfoundland and the Azores, but we have hopes that we can raise her. . . . The minutes pass. No good. She is too far away.

We raise New York at 1920 and give our position. We try Santa Maria too, for the first time—and here is Santa Maria, loud and clear. We are now about five hours out. The time has just sped by; it hardly seems possible that we have been going that long. We have Argentia, Newfoundland, on the ADF, bearing about 280 degrees. We are moving slowly toward land's end in North America; soon we will have these friendly beacons no more. As far away as they are, they are a link nonetheless, like milestones on the road, towns and villages which we pass unseen and unheard, but which are as real to us out here over the ocean as though we could in fact see their houses, churches, stores sliding past in the darkness of distance, brief glimpses of light in the long vista of our journey.

There is another ship, seen through a great hole in the cloud floor to our left. The third one so far; tiny, remote—so far below. It would be nice if we could talk to it.

At 1950 hours we try *Ocean Station Delta* once more. The ADF shows a weak signal; Max can't improve on it so we let it go. In the past, on occasion, he has asked for a continuous beacon to navigate by; this time we don't need it, at least not yet.

2005 hours, and I see a strange and beautiful rainbow effect on the thick, gray, endless floor of clouds ahead of us—a pastel-colored rainbow, an elongated elliptical pattern of lavender, purple, pale red, pale blue. On this silent landscape of great, piled-up layers of clouds it is lovely. Is it ice crystals, perhaps?

The skyscape is immense, darkling, limitless, and lonely. The sky is slowly turning a deep blue-purple overhead, the sun's rays are weakening behind us and I have a strong feeling of night coming, and cold. It is one of those rare (so far) little moments of apprehension, when I realize with a brief stab of clarity where we are, how small we are in this vast air-water world, how vulnerable we are. I don't allow the thought to grow, and there is no reason to. We fly untouched by any of these risks and perils, two engines which have never even had a catch in their beautiful and reassuring rhythm, a warm cabin isolated against the coming night, a friend who sits now hunched over his radio, the headset incongruously framing his unruly gray hair, glasses perched on his nose, his face intent on the little, red-glowing windows of the ADF set which reaches out with unseen voice and ears to bring signs of other men, friends of passage, far away in the darkening world outside.

Yet the sense of loneliness at this moment is inescapable and inevitable, for we are flying in an utterly fantastic world, a world so vast and strange that in it we seem the merest microcosms, aliens tolerated by the silence because of our total insignificance in its limitless expanse. It is an enormous cold gray world of huge and tumbled clouds reaching as far as the eye can see beneath the chill dark sky. In this lonely landscape now and then a pool appears, a pond or a lake—

openings of various sizes from small to miles across like pools in a landscape of unending snow. And now a thin, thin veil is drawn across these openings, the finest of gauze, and the almost forgotten, final rays of sunlight make beams of light across the peaks and valleys of cloud. In this veil the pastel rainbow appears and disappears, shimmering like a pale will-o'-the-wisp, rising and falling across the hills and gorges. And every now and then, through some such lordly chasm, the sea appears below, cold, cruel, wind-whipped, and sable-gray.

2020 hours. Almost six hours out of Boston, and Argentia is abeam, a faint but still musical signal bearing 270 degrees. We tune in Torbay, land's end for us on the tip of Newfoundland near St. Johns; it comes in even more faintly, and almost on the same bearing as Argentia. There is perhaps a spread of 12 degrees between them; projecting this on our weather chart, we find we are somewhat further south than we had thought. We correct our heading a bit, and steer 115 degrees. We are losing our contact with the land.

2045 hours. The sun is momentarily lost behind the clouds which rise almost to our height now.

2045 hours. We call New York, get no reply. We call Santa Maria. Silence. We call again: "Santa Maria, Santa Maria Approach, this is Apache Delta, Gulf, Alpha, Romeo, Yankee . . ." We get an answer loud and clear, from Gander. Max gives Gander our position and our estimate for our next call at 2215.

The sun sinks lower, and our pastel rainbow grows in size, diminishes in light, moves ahead to the horizon and finally disappears, merged into the gray and lavender of the eastern sky. Behind us the sun goes down in a blaze of red and orange. Outside temperature is 15 degrees, air speed a steady 130, power settings unchanged. The light grows dim; soon it will be dark. The moon, too, has moved behind us; it will set not long after the sun and we will be alone with the stars.

2100 hours. The sun has disappeared.

2105 hours, and Max tries *Ocean Station Delta* on the ADF again. The needle circles slowly around the dial, stops at around 80 degrees and pulses restlessly and rather vaguely. "She's hunting for something there," says Max "and that's about where the weather ship ought to be." To check, he switches it off briefly; the needle turns on the dial, but comes back to the same heading when he switches on again. He tunes in on the Torbay frequency; slowly the needle homes on 2 degrees. Both stations are too weak to hear; both indicate by their weakness that we may be farther south than we had thought. But that's okay; we'll be blown back northward soon enough when we reach the other side of this counterclockwise wind pattern.

Darkness is closing in fast now. Outside everything is pearly gray; the sky behind has changed to rose. The deep chasms in the cloud floor seem mysterious and vast; dark, silent rivers winding though the landscape of an unreal world.

2152 hours. We call *Ocean Station Delta.* There is no reply, but its signal is coming in strongly now. Max pulls out his facilities chart and finds the weather ship on it—it patrols an area marked off in squares, each square lettered across and down. By the letters in its signal we can tell where it is: if it sends C and P at the end of its regular call letters, for instance, we find C on the grid, run our finger down to where it meets row P, and there is where it is.

A curious thing develops. We seem to be getting two different signals from the weather ship—both with its regular call letters, but with different location letters. One indicates it is in the southwest corner of the grid area, the other that it is further up and over on the eastern side. We check and recheck. Are there two weather ships in the area? We don't know.

2215 hours. We call Santa Maria and give our position. We also start up to 9,000 feet and ask for clearance to that altitude. The cloud floor below has been rising slowly to meet us as we travel onward; now we are beginning to brush the peaks of that unearthly landscape, and little wisps of fog make brief flashes of our red and green navigation lights on the wings as they whip past. It is an eerie feeling, particularly when we cross one of those cavernous openings that lead straight down to the dark sea so far below.

We send our request for a new altitude again and again to Santa Maria, without reply. Meanwhile I milk D-GARY slowly upward, out of the reach of those looming, seemingly so solid peaks of cloud.

2235 hours. Santa Maria comes in and clears us to 9,000, but wants us to let down to 7,000 again when we reach 40 degrees west. We're not far from there now, but let's enjoy our higher level while we can.

The weather ship, *Ocean Station Delta,* is coming in strong now, and we are almost abeam of it. Still no solution to the mystery of the two divergent signals. Max thinks that perhaps the relief ship is coming in. The moon is low and far behind us now. The stars are clear and glittering in the cold blue sky.

2258 hours. We pick up the beacon at Lajes, the U.S. base in the Azores, on the ADF. The signal is faint but the direction is positive, and we are exactly on course.

2305 hours. *Ocean Station Delta* is transmitting again, but again the signal has changed. We check its position on the grid—much too far for it to have traveled since we last heard it. A mystery. But again we are drawing abeam of it, almost have it, in fact.

2320 hours. We are abeam of *Ocean Station Delta.* The clouds are piling up now in great soaring peaks, reaching almost to our new altitude; the floor is breaking up, now the skyscape looks like a series of vast, scattered mountain ranges with great, dark open spaces in between. Our cabin is bathed in the warm red glow of the instrument lights which shine down from the ceiling onto the panel and make the radium dials glow.

We call *Ocean Station Delta,* and receive a strong answer from MATS 3306, a Military Air Transport Service plane, asking if we need any assistance, sir. We are flattered by this form of address and ask him for his weather, whether he has heard from Santa Maria and if so how he reads them. He replies that he is two hours out of Lajes and that the weather for Lajes is expected to stay good. We ask him what his altitude is. "We're at fourteen thousand in the clouds," he says, and it develops that there is a front between us and the islands but that once through it we should be in the clear. "By the way," says Max in that low, slow, careful voice of his, "we were talking to somebody called McKinley, it sounded like, some time ago—do you know where or who McKinley is?" Back comes the strong voice from MATS 3306: "Sounds like Bermuda to me, Kindley Field in Bermuda." We thank him and sign off, and MATS 3306 wings on his way. A friendly fellow, and it is a comfort to think of him, big and powerful, 14,000 feet up and homeward bound.

2340 hours. We call Santa Maria to give our position, and Clipper 154 answers. He sounds crisp and efficient as all hell. He relays our Santa Maria call, then comes back with the request that we try another frequency. Okay, say we, and ask where he is. He gives his position—behind us quite a way, at 21,000 feet. "He'll be catching us and passing us before long," says Max, "and we probably won't even know it."

We are drawing near that front now, and a few small bumps remind us that it is time to tighten our seat belts. I peer ahead into the darkness—a darkness dimly but surprisingly well lit only by the stars. It is a diffused, all-pervading sort of dimness, less a light than a lightening of what otherwise would be total blackness—and by it I see the front ahead, a great, dark towering bank of cloud reaching right across the horizon, a looming cliff of blackness with a grandfather's fringe of tufted white along its top, blowing with the wind.

If MATS 3306 is at 14,000 feet and still in that, we can't possibly climb over it. So it's straight ahead and in.

2350 hours. The temperature is rising—it's above freezing outside now. We have entered between two layers of cloud, one below, one blotting out the sky above. The front still looms ahead—it is awe-inspiring, like flying almost blindly toward a mountain range that we know we cannot climb.

0046 hours. Nearly an hour has passed. We are in it now. The layers of cloud closed gradually, the clifflike wall of the front grew dark and dim, and then with a rush there was cloud all around. It is raining hard. The green light flashes on the wingtip at my side as fog and rain tear past it. D-GARY leaps and drops, bucking. The rain makes a rushing, peppery noise in our dim little cabin, a steady, sharply sibilant stream of sound to match the streaming water on the windows. Max pulls on the carburetor heat; we are getting a little ice there. I hold the flashlight on the carburetor air heat dial while he slowly, with infinite care, adjusts the knobs until both dials read 100 degrees. Once before, on his first trans-Atlantic flight in an Apache, he held those knobs for thirteen hours with one hand while flying and tuning his radio with the other—thirteen hours of storm in which he had to hold them because otherwise they slipped back in, the jets iced up, and his engines started to cut out.

Now it is rain and snow mixed. The sharp sibilance becomes an incredibly fast, rattling tattoo. Max switches on the taxi light in the nose—it is an amazing sight! Snow streams at us; a whole world of snow is aimed directly at this little airplane, a million lancelike streaks of its whizzing at us out of the blackness, stabbing at the windshield, a world alive with hissing, rattling, streaming snow. He turns the light out, and we are in blackness again, but a blackness still alive with sound.

And now it is alive with light again, but a different light—an eerie, flickering blue that runs up and down the windshield frame, leaps off to the wings, appears again, flowing in jagged edges along the plexiglass of windshield and window.

"Static electricity," Max shouts above the noise. "Watch!" He puts his hand up to the windshield, touches it and jumps it back with a grimace as a long blue spark leaps from it. I reach forward tentatively and stare in astonishment: long blue sparks trail from my fingers with a prickling sensation, a stream of them from each fingertip when I approach the plexiglass. I am too fascinated to be even frightened.

We fly on through the storm. Max tries to tune the ADF; it is no use—it points only straight ahead, drawn by the static charge in the storm outside. The constant stream of sound ebbs and flows as the snow waxes and wanes. We fly on, cut off, alone. There is no fear, there is only a hypnotic fascination as if I, too, were being drawn irresistibly by the sound and fury toward the heart of the witching world in this great cloud of storm.

0110 hours. We are out of it. We have been in it a bare half-hour. It seemed an eternity. And then it stopped, as suddenly, more suddenly, than it began, and we are abruptly in the clear.

In the clear, but in a skyscape more awe-inspiring, more immense than anything I have ever imagined. Ahead and on both sides, great towering masses of cloud loom into the black sky. There is no light, only the ghostly dimness of stars somewhere far above, yet everywhere these tremendous soaring peaks are visible, ghostlike, a fantastic, dreamlike scene. We seem to travel in a vast stillness in which nothing moves except ourselves, a tiny errant star wandering forever in a place lost to time. And there, like a companion star appearing on our right and far above us, I see a plane.

I have a vague idea now what a man in space would feel

like when he saw another man. I strain my eyes toward those blinking lights, one red, one white, that hang suspended far above us, and seem to see into the cabin of that companion ship, with its comfortable, reclining seats, its soft lights, its sleeping people, its stewardesses walking quietly up and down the aisle. It seems so near! We ought to get together, you and we, I think; we ought to sail the rest of this way wing to wing, for this is a wide, wide world and we are very much alone. But he draws ahead, wings on and slowly disappears while I stare after him; at last I see his lights blink one more time and he is gone.

0145 hours. Rain, hard and short. Then we are out of it and flying between layers of cloud. We see another storm to the right and turn to skirt the edge of it.

0155 hours. A sudden whoosh from the right engine; the wing dips sharply; we are out of gas on the cabin tanks. Max reaches down and switches on the cross-feed from the auxiliaries; the engine picks up and we fly straight and level again. I unplug the cigarette lighter. We fly on, pointing for the Lajes beacon. It is time to look for lights now; we should be nearing Horta.

0205 hours. Calling Santa Maria. Max, with his earphones and glasses, bending over his charts and maps, speaking into the mike, has the look of a pixyish scholar, incongruous up here in the darkness and the clouds.

Santa Maria, Santa Maria, Romeo Yankee calling. No reply.

Santa Maria, Santa Maria—and now I can see he has an answer. But he's mixed up; he keeps saying, "Gander." "I read you, Gander, five by five," he says. Does he think he's up north? No, it's Gander, all right—and Gander relays us to New York, and New York relays us to Santa Maria, and we are advised to try Santa Maria on another frequency, 5626.5. All of a sudden we are not alone any more—the sky seems positively crowded.

0220 hours. We call Santa Maria on 5626.5 and get New York. What an odd thing this radio is! We give New York our position, and then we get Santa Maria, too.

0228 hours. Max sights a beacon on the right, a light! It's a light for sure, not a star—a light down there in the darkness, probably a buoy, rocking on the ocean swells far, far below. The radio compass has started to swing to the right now too; we must be almost abeam of Flores, the first of the island chain.

Now begins a period of curious anxiety for me. I have been watching the various dials before me, and I note with a start that the tank gauges show only about half full. We are not on the cabin tanks any more; does this mean that half of our remaining gas supply is gone? I check several times in the next twenty minutes or half-hour, and the gauges seem to be going down with startling rapidity. We have been studying Max's chart of the islands; I know there is still a long way to go. Even when we have come abeam of Horta, with its 7,000-foot mountain (which is on our minds, too, Santa Maria having brought us down to 7,000 feet), we still have about 150 miles, down past Lajes and Santa Ana to Santa Maria, at the bottom of the island group. I begin to watch those gauges with real concern; I don't know whether they show all our remaining gas supply, and I hesitate to ask.

We see stars now, behind a thin veil of high cloud. Cloud

mountains still loom all around us, and once in a while we pass through a brief but violent squall. Below, the cloud floor is broken. We are tuned to Lajes beacon, and the signal is a steady, comforting sound: .... —— .—. ———, ending in a long steady musical tone.

0315 hours. A light beacon dead ahead. Then another to the right, and still another. There is a ship, too. We are approaching an island; it looks like an island, dead ahead—a long sliver of dark land sloping down from a mass of land to the left. We approach it, and it seems to melt away into the darkness. Was it an island? I will never know.

0450 hours. Horta is abeam. And at its last moment of life the moon has reappeared, to set behind us in a pale glow of moonlit cloud. I had thought it long since gone.

0455 hours. We have climbed to 9,000, thinking of that mountain, and now we call Santa Maria to ask for the winds at 7,000 and for Santa Maria's weather. The mountain is at last abeam of us, and I see it clearly—very solid land indeed, a great, dark bulk, almost as high as we, a sharp cone of a mountain with a long island spine and ripples of dim white cloud climbing up its far side.

0415 hours. We tune in Lajes omni and check it against the ADF. We are coming abreast of Lajes now. About 150 miles to go.

We have strong head winds. Santa Maria has sent us up to 9,000 feet again after we started descending, right where the winds are strongest. They explain this long ride down to the islands and along the chain: we were blown further north than we thought and have had to fight our way southward since the storm.

0515 hours. We have flown for a long time in silence, in and out of clouds, watching, waiting. Now we contact Santa Maria tower. We are still about 50 miles out, with Santa Ana abeam. Santa Maria checks our position and heading, gives us altimeter setting and winds.

0525 hours. We have burst out of a cloud and there, as abrupt and as beautiful as the curtain-raising in a darkened theater, we see bright lights of the airport—twin yellow lines of the runway, dim blue glow of the taxi strips, a beacon flashing, lights moving along a road, the warm glow of a town. As quickly as we see it, it is gone, and we are in a cloud again, but we turn in a wide, descending spiral and there it is once more. The sight is unforgettable. We are poised between great pillars and mountains of cloud, threading our way through measureless chasms that wind between them, circling, turning, descending in breath-taking sweeps. Santa Maria is talking to us over the loudspeaker now, giving us instructions, asking questions, getting answers. Romeo Yankee is reaching port and the pilot is guiding us in, in strongly accented but very clear English: "Romeo Yankee, descend to three thousand feet over the beacon, fly over it on heading 190 degrees, make a 180-degree turn . . ." We briefly flash into a cloud again, out of it, we sweep down, the runway is ahead. Max dumps lift with the flaps—once, twice, three times—the gear comes down, the runway lights flatten out, loom large, start flashing by. D-GARY, alias Romeo Yankee, touches with a soft thump-thump and we are on the ground at 0539, one minute before Max's estimated time of arrival.

The tower guides us down the runway to the Follow-Me jeep. We speed along behind it, see a figure waving lights, turn, slow, stop. Max cuts the engines and we climb stiff-legged from the plane. The air is soft and warm. A huge Lockheed Constellation radar picket plane stands hulking in the blue luminescence of the taxi strips, busy figures squatting around its pregnant belly. We hear American voices from its crew. Ahead, the lights of the administration building beckon and we walk in.

Later, over a quick breakfast in the empty, silent dining room of the airport, I ask Max about those fuel gauges. "Oh," he says, "those were only the auxiliaries. We still have about five hours left in the mains." I might have known.

We are in Santa Maria about three and a half hours. That's time enough to check in and out of Customs (no problem, just filing some papers), file a new flight plan, get a new weather briefing, get all tanks filled and the airplane checked again (after all, there are still some 800 miles of ocean to go, though one feels inclined to minimize the distance after that long first leg), get breakfast all alone in the cavernously empty big dining hall (ham and eggs for me, with coffee; toast and honey for Max, with milk), and make good use of a land-based toilet (a couple of empty paint cans served us in the air). Everybody at Santa Maria is exceedingly friendly and helpful. The Constellation contingent, a bunch of crew cuts in Navy suntans and leather jackets, is aloof, incredulous, and inclined slightly to look down their noses at us; but after all, they have their own problems, and shortly after we land they take their air-borne electronics laboratory and roar the hell out of there, Malta-bound. We crank up

Romeo Yankee just after daybreak and I take her off into a cloudy sky at 0932, Greenwich time.

At 4000 feet Max suddenly takes over, turns, and we go whooshing back down again. He has forgotten his briefcase. Santa Maria advises that the jeep will be on the runway waiting with it. We lose altitude in a long, shallow dive, passing so closely over the rugged coastal hills that I can clearly make out the tiny white, red-roofed houses, the hilltop chapels, the brown roads. We skirt the airfield, drop down almost to sea level, make a wide turn and come in low, on a direct line with the runway, level with the cliff on which it is built. We rise just slightly as we reach the cliff's edge, spill flaps swiftly, and touch down, full tanks and all, with scarcely a jar. This man can really fly!

The briefcase aboard, I take off again and start the long, slow climb to altitude. We are cleared to Delta exit point, about ninety miles out, thence VFR to Madrid at 11,000 feet. The weather reports are fine for this leg, but less promising for the next day—Paris is weathered in with fog, Düsseldorf too, Geneva is not promising. But here we have a ceiling of 1,500 feet, broken clouds, the same towering, majestic formations which awed me so last night, but with blue skies above and steadily widening gaps as we climb out on a heading of 110 degrees, almost directly into the sun.

Max has told me to climb at 120 miles an hour, no more and no less, and he can be strict about this sort of thing. I keep my eyes glued to the air speed. This is work. And then, quite suddenly, I am also fighting against sleep. The sun in my eyes, the warmth of the cabin, the drone of the engines all combine to put me in a hypnotic state. . . . Suddenly I am far, far away, I hear nothing, feel nothing, see nothing; my hands are on the wheel, my eyes are open, but my mind has fled to distant places . . . dimly I feel the airplane climbing . . . something is wrong . . . some drugged sense tells me that we're climbing too steeply. . . . I wrench myself awake with a fearful start. . . . Holy smoke, the air speed! It's down to 110 and going lower. . . . I shake my head to clear it, push the wheel forward, concentrate. . . .

How does Max fight this sort of thing on a long flight, a really long flight? It is insidious and dreadful; it comes without my even being aware of its coming. I am flying, I am concentrating—suddenly I am gone, and only that deep, deep sense that doesn't quite go out saves me, in the nick of time: I feel the climb steepen and somehow jerk myself awake. But suppose I didn't feel it? Is there a time when even the deepest senses are overwhelmed by sleep? I try looking out the window, scanning the skies, varying the routine. It helps somewhat. I remember what Max said about keeping himself busy all the time. I can see now that with someone else along, and the opportunity to relax, he is giving in to the drowsiness too—he nods from time to time, his face, lined, stubbled, gray, suddenly dissolves in brief sleep with a look almost of pain.

0954—On top at 7,000.

1004—We have climbed out at 11,000 feet. CAVU with vast, tumbled floor of cloud ahead. We estimate we are at Delta exit point—how long ago it seems since we left EEL, out in the ocean east of Boston! And the ocean is still our world; it is hard to believe that in another five hours or so, another 700-odd miles, the ocean will give way to real land, Europe. . . .

1040 hours. Cruising at 11,000 feet. I am trying to trim her out for level flight, but at this altitude and with this load it's like trying to balance her on a log. No sooner is she up there than she starts sliding off again. It's a real job. Air speed is a bare 120 mph and she just barely holds her altitude at that.

It is a majestic day. We have left the cloud floor above the Azores behind, and we now have an unobstructed view of the sea beneath—deep blue, immense, crinkled with waves. Far ahead, reaching into infinite distance on both sides, is a tremendous bank of clouds—magnificent great cloud ranges, tier on tier, peaks and plateaus, silent, unmoving, remote, seemingly as solid as a snow-covered earth itself. The sun shines bright and hot through the windshield. The outside temperature is just about at freezing. The plane moves quietly along, detached from every worldly thing. Santa Maria asked us to stand by nearly an hour ago, but we have heard nothing since and so we shut the radio down.

1050 hours. Among the peaks, a glacier landscape looms ahead, a Greenland of the skies.

1103 hours. We report our position to Santa Maria; it's sort of nice to hear that familiar voice break into our primeval stillness up here. We are now almost past that tremendous mountain range of clouds; it lies below us like a heavenly Alps, and beyond is a Sea of Azov, veiled and mysterious, island-dotted, the way I would imagine the Sea of Azov would look, or Alph, the sacred river, in Coleridge's "Kubla Khan"—and beyond it looms another distant range.

Behind us now is a most astonishing sight; clouds that seem to be reflected in a glassy lake, like a mirage. It is a really surprising illusion, as though at this altitude there were another little sea above the sea, mirror-smooth. It is caused, I think, by a thin layer of broken clouds, lying between the big layers which look like white islands on the glass-clear water.

1145 hours. We have crossed our mysterious Sea of Azov and are reaching the other side; and it lost none of its mirage-like qualities in the crossing. Now out of a platform of cloud ahead mushrooms have begun to sprout by the hundreds. Some are slender pillars, some bulbous-shaped, some squat growths—the cloudscape is dotted with them as far as the eye can see. Warm air is rising from the ocean, pushing up the cloud layer; these are thunderheads aborning, baby cut-ups that can combine and grow to huge and violent size. It is almost like watching the process of Creation.

1205 hours. We contact Santa Maria again.

Once more I am getting very sleepy when I am at the controls. Max thinks that the altitude has something to do with it, a touch of anoxia. I get sudden, sharp hallucinations, snap out of them with a real physical jerk. I have an idea. I remember the Wash 'n Dries I brought along, get out the box and tear one open. The little paper towel, saturated in some astringent liquid, is marvelously cool and refreshing. I go over my whole face with it, down into my neck, and the lively, pleasant smell of it fills the little cabin. Max is impressed; he tries one too.

1305 hours. I've had an orange and some cold chicken, then some coffee and I feel better, less sleepy. After a short harmonica session we call Santa Maria again and give our position. We're well over halfway now.

1338 hours. We try calling Lisbon approach on various VHF frequencies. No reply, still too far out.

The sky is dirtying up to the southward now; out of my window I can see a high, murky overcast with rain squalls here and there. We are just going to pass the northern edge of this. Straight ahead and to the north it is CAVU as far as the eye can see.

1411 hours. We are hearing all sorts of airplanes talking to Lisbon approach now, but we don't have them yet. That means we are still at least a hundred miles away. We try again, and now we barely begin to hear Lisbon approach answering the other aircraft. It is a weird feeling: we will hear some pilot somewhere giving his position, asking for instructions; he breaks off; silence; then he comes back and acknowledges a voice we didn't hear at all. But now, ever so faintly, we hear the voice replying, and we now are getting close to land.

We give our last report to Santa Maria now. We are nearly five hours out, and it is "Thank you for everything, Santa Maria, and good-by." I remember those calls of the night before, in the dark and the storm, the instructions relayed from New York and Gander and Shannon and Clipper 154; and the warm and beautiful sight of that airport suddenly appearing from behind the curtain of cloud. "Good-by, Santa Maria, and thanks—I'll never forget you"—but look who's saying good-by without a radio.

Minutes later it's "Hello Lisbon," and the new contact is made.

1420 hours. Lisbon clears us down to 9,000.

1430 hours. Land ho! Frankly I don't see it yet but Max says he does. We are 24 hours 6 minutes out of Boston via Santa Maria, Azores. We reel up the HF antenna and put the earplugs away. No need for the long-range set anymore.

1440 hours. We simultaneously acknowledge that for some time now we have been hearing a peculiar buzzing sound from the right engine. It sounds to me like a vibration somewhere. Max turns off all the radios to see if it is a radio hum; no. He checks the landing gear; no. We don't know what it is. It sounds like a loose cowling, perhaps; some thin piece of metal vibrating in a hornetlike buzz. But it doesn't seem to be anything serious; all the instruments read okay and there is nothing loose that we can see.

1505 hours. That land was a false alarm. We must be quite a way out still. We try a VHF contact again; no go.

1512 hours. Land ho! This time for sure. At the same time we get Lisbon on VHF. Max tries Lisbon omni; the needle swings and he centers it, so now we have omni and the ADF. What a wonderful feeling!

1530 hours. The land is clearly in sight now. I see the sandy banks of the Tagus River that I remember so well from my visits as a foreign correspondent years ago. We are coming right up to the mouth of the Tagus. Fishing boats appear below, and an occasional freighter. It is CAVU all around, a beautiful, sunny sky. The air waves are full of a mixture of accents—American, British, French, Spanish: this is Europe again, we have made it. "We could swim from here now," says Max with a smile. We could, we could indeed. There is Escorial ahead, and the shore road to Lisbon;

Cascais, the strange monastery crowning the black mountain, with its forbidding brotherhood of stern and silent monks; and the smoke of Lisbon. We are to report at the Lima Sugar beacon, and we start watching the needle for that now. "They're real strict around here," says Max. "We want to be sure we hit those beacons right."

1557 hours. Lima Sugar beacon passes below, and we are cleared to Charlie Romeo at Coruche. There's the airport; we fly right over it at 9,000 feet precisely.

1612 hours. Charlie Romeo below (we see the beacon this time); Lisbon passes us on to beacon Papa Mike at Portalegre. Gone is the ranging freedom of the sea; we are controlled like traffic on a highway here. We estimate Papa Mike at 1644.

1646 hours. The needle wavers; Papa Mike is below, and we are cleared to Charlie Charlie, which seems to be at the border. Max is flying with his facilities chart spread open on his knees now. We have to watch and plan and estimate.

We are never sure just when we passed Charlie Charlie. We couldn't seem to make it out, but for documentary purposes we estimate it at 1724.

I am flying now in the peace of the afternoon. For all the dreaming I have done since I was a small boy, I have never known that flight could be so beautiful. We are at 9,000 feet over Spain. Below is a broad river valley, a river that must rise high at times, leaving wide expanses of sand when it is low, and bordered by fields of an incredible green. Rolling hills rise from the valley, carefully tended, with fields that look, from this height, as though they had been combed. Here and there are towns and villages nestled in the soft curves of the earth as though, after long wandering, they had found shelter there, liked the place, and stayed. Smoke rises softly in air that is almost still. We are trimmed out in perfect level flight; it needs only the gentlest touch, not more than the crooking of a finger, to keep the Apache precisely where I want it to be. The engines hum. They are so perfectly adjusted for our slow, peaceful pace that they are no more obtrusive than the noise of an automobile engine running quietly along a wide, straight prairie road on a summer day. The air speed needle is steady on 130 mph, the ADF points straight ahead, the altimeter hangs unvaryingly around 9,000. And then, off to the left in the distance, I see the shadowy cloud of a snow-capped mountain range.

We fly along those mountains all afternoon while the day slowly fades below. Slowly, too, the country changes. As the mountains rise, the green below gives way to browns and reds. The hills grow sharper, the broad valley splits, its tributary valleys narrow to gullies, deep, steep, eroded. From red the dominant color fades to gray and black, the landscape turns harsh, tumbled, arid. Ahead of us appears a long, low, dike-like formation, so regular, so flat-topped that it seems it must be man-made—except that it is probably at least a thousand feet in height. This, too, splits, splits again and I see the fingers of black foothills rising toward another range on our right, a range just powdered with a sugarcoat of snow.

The stillness is immense and beautiful. For probably two hours neither of us has said a word; we sit, lost in our thoughts, and my fingers gently guide this lovely plane. I sense that the sun is setting behind me; on the ground the shadows grow long, and haze draws across the strange and unreal landscape like an evening shroud. Now the snow-capped range at the left—the Guadarrama Mountains, I later learn—turns pink, then red, then gold. The land grows dim, then dark; at 1800 hours the sun sets over Spain. Twilight enters our little cabin, too; but now the spell is broken; it is time to work again. Ahead, a soft glow in a hazy, slate-blue sky, are the lights of Madrid.

We come up on them swiftly, and as swiftly as the night flees over the countryside below, as swiftly is the countryside transformed. Lights wink and blink everywhere. Houses give way to groups of houses; lone cars on the highway become many cars, their headlights wavering along, flashing briefly as they top a rise and shine into the sky. The glow of Madrid becomes a blaze. Our cabin light has been turned on and Max studies his facilities chart. Then he speaks quietly into the microphone: "Madrid tower, Apache Delta Gary Alpha Romeo Yankee—Romeo Yankee calling." Madrid replies, and we are cleared to Runway 33. Max motions, and I start to descend.

At 2,500 feet he takes over and we soar across Madrid's outlying hills. I see the city blazing to our left, a riot of gold. Lower we come, and lower. A hill looms ahead, close, black, much higher, it seems, than we are. For the first time in this entire journey I feel a real stab of fear: Max, his eyes intent on the darkness ahead, doesn't seem to have noticed it. The hill rises gigantically—my God, I think, we're going to hit it. And suddenly it is gone, it flows past below, we stab through a brief, thin wisp of cloud. "Good Lord," I say, "Max, I thought you were going to hit that hill for sure." He smiles. "It only looks that way," he says. "I'll show you tomorrow."

He speaks into the mike again. Straight ahead, the twin lights of a runway flicker; then, at almost the same instant, another springs into view at right angles, to the left of us. For a moment we are confused. Max checks the tower: "Romeo Yankee calling, which runway are we cleared for?" "Romeo Yankee," comes the answer, "you are cleared for Runway 33." Our heading is still east. "Oh, yes," says Max, "I remember, that's a military field ahead. We don't want that one." We swing to the left, line up for the twin paths of gold that shine beyond a low hill. Down, down we come. We drop the gear; the three lights flash on, indicating that our wheels are down. We bank gently, lining up. Whoosh! go the flaps—we balloon and slow. The concrete ribbons past, the numbers 33, streaks of tires. We spill lift with the flaps, drop gently down, touch and roll. The blue lights of the taxi strip appear; the tower guides us in. It is 1843, Greenwich time.

Twenty-four hours later we are peering down through a veil of fog at Le Bourget in Paris. In 6 hours and 20 minutes we have flown across Spain, climbed the Pyrenees, winged past Biarritz (the Atlantic, one more time, a blue expanse in the distance), past a giant layer of smoke that covers Bordeaux where a huge brush fire is burning, up the valley of the Loire to the flat floor of fog that spreads out over Paris. It is evening now, the sun is setting, and a jet leaves a fiery contrail in its dying rays. Our cabin tanks are empty; we drained them as we approached Poitiers, flying nose high, right wing low, like a tilted drunk until the engines whooshed and died and we knew the tanks were dry. We have seen the Eiffel Tower sticking like a finger through the fog that covers Paris, its skeleton visible right down to the first balcony. We have flown around the west side of the city looking for Le Bourget, and there it is below. Hangars and runways are clearly visible; yet we know that down there visibility is barely half a mile and the ceiling two hundred feet or less.

Düsseldorf is fogged in; we have checked. Geneva was no better today than it was yesterday. We debated various places; in the end Max said: "We'd better try Le Bourget; they can always talk us down."

They are talking now, and we reply. "Le Bourget, this is Romeo Yankee. We are over the field." We get instructions how to proceed; let down to the beacon, overfly it at 2,500, Ground Control Approach will take us from there.

We go down, and gradually the haze thickens. As we approach it more and more horizontally it is no longer a veil, but gradually becomes a wall. The hazy outlines of the earth disappear. We are heading toward what seems to be an impenetrable floor.

Now GCA takes over. "Romeo Yankee, check your directional gyro and give me your heading." We do so, synchronizing our gyro with our compass, and report our heading as 269 degrees. "Romeo Yankee," he replies, "turn one degree to the left. You are four kilometers out, your heading is 267." We confirm each heading as he gives it to us. "Hold that heading," he continues. "Your glide path is good. You are four kilometers out, your heading is 267.

Correct your heading to 268. Three kilometers out, 268. From now on do not report back. Correct to 269. Two kilometers, 269. You are over the glide path, 269. Your glide is good. One kilometer . . ." The glide path, a shimmering line of gold, suddenly appears ahead, flows past below. Runway lights glow. "One kilometer, 269 . . ." There is the runway. We touch in a world of murky gray at 1820, and GCA turns us over to the tower frequency of 119.1. The trip is over.

The next day, in a cavernous hangar at Le Bourget, we shake hands and say good-by. Romeo Yankee is all cleaned out and spick and span. She will stay there until her owners fetch her. Max takes a plane the next morning for New York; I take the night train to London.

9. *All on My Own*

Three thousand feet above the green and brown of Martha's Vineyard a little airplane turns and twinkles in the sky. In the clear air, with the sun's light glancing off its wings, it looks like a bit of heavenly tinfoil drifting down out of the blue, rising, falling, tipping a wing to slant off in a swift drop to one side, recovering again with a distant burst of engine noise. Now it noses up sharply, turns, falls in a spiral, straightens, dives, and rises again to climb the invisible heights. Below, on the vivid blue sea, its shadow chases over little crinkled whitecaps and my wife, lying on a deck chair beside the swimming pool, wonders: "What's he up to now?"

The cabin of the little airplane is a busy place indeed. I sit at the controls, doing things with wheel and rudder and throttle that I had almost forgotten. Steve Gentle, at my side, is calling the turn on me. "Level her off, now. Okay, okay. Cut your throttle. Bring your nose up. This is an approach stall, see? You're coming in for a landing. Get your nose high. Pull first flaps. Bring her back, back! Get the nose higher! I want a nice clean stall now, and a fast recovery. Bring your wheel back, back!" And as I pull back on the wheel and the nose rises higher and higher into the unfamiliar, uncomfortable, unnatural position of a full stall I am already thinking ahead: when she breaks, shoot power to her, get your nose down, get air speed fast before the nose falls below the horizon. . . . A nice, clean stall. . . . Now! Now! And the motor roars in a swift crescendo, and we drop, and the air speed needle, which has fallen to below 50 miles per hour, starts forward on the dial, toward 60, 70, toward safety. . . . I have wind on my wings again; I can fly.

Thus it goes, through one maneuver after another, hour after hour, day after shining day. Steve Gentle belies his name; he is a hard taskmaster to me in the air. I have come up here with more than twelve hours of solo cross-country in my logbook, feeling myself a seasoned student pilot ready to go at the private pilot's flight test. I am flying 9013-Delta, too, the ship I soloed in, my old love. She takes off like a bird, and I can land her on a dime, I am getting impatient to share my skills. I am awearying of flying alone, I want to take my wife and children up and go places and do things. . . .

And here I am, doing them. Only not the things I yearned for. I have forgotten how much I still need to know. And at times it seems almost discouraging.

First, stalls. I have to do stalls with power on and power off, stalls in climbing turns left and right, stalls in descending turns. Stalls in a modern small plane like the Tri-Pacer are not too difficult to execute, nor are they dangerous, but they are uncomfortable and unnatural. I do not like them, but I have to do them again and again—because what is difficult, even delicate, is to do a clean stall and recover from it well.

A stall in a climbing turn, for example. This is the one that kills pilots close to the ground. It is the type of stall that can be the fatal climax of a careless take-off—the plane leaves the ground, climbs out, engine at full throttle, nose perhaps a little too high, air speed a little too low. The pilot starts his left turn in the standard take-off pattern. His nose drops a bit as his wings bank; he pulls the wheel back to lift it. Unnoticed, his air speed falls off still more. In the banked position, the load on his wings increases; the wings are fighting not only to maintain lift against gravity, but against the pull of centrifugal force as well. Instead of stalling at 50 miles per hour, the plane may stall at 60, or 65, or even 70, depending on the steepness of his turn. And now, as the pilot pulls the wheel back to bring his nose up, the fatal sequence has been launched: the air speed drops still further, the plane labors, the nose begins to drop again, he pulls back on the wheel still more—and stalls. At 300- or 400-feet altitude, in that situation, it is more than likely too late to recover.

How does it feel to have this happen? Three thousand feet high over the Vineyard, Steve shows me.

We simulate a take-off: I give the Tri-Pacer full throttle, as though we were rolling along the runway, then gradually ease her into a climb. Higher now, and higher still. The horizon drops right out of sight; we are pointing eerily up into the blue sky. Now I start my turn. Ease her over to the left, just as though I were in the traffic pattern. The nose falls off; bring back the wheel still more. I steal a quick glance at the air speed and am startled: it is down to below 70. I am turning now, and I bring the wheel still further back; I can feel 13-Delta laboring, protesting; we seem to stand on one wing. "Now!" says Steve. "Bring her all the way back!" And I haul on the wheel and she shudders, makes one more effort, gives up, swoops off down and sideways. The sweat is rolling off me; in that position it seemed she might have done almost anything; but at the stall I push the wheel forward to give her flying speed, and in an instant she is out of it and we are diving cleanly; I bring the wheel back slowly and now we are straight and level once more.

I do this to the left, I do it to the right, and I hate myself every time I do it. It is like forcing my beautiful ship beyond her strength; it seems a brutal maneuver. But the signals that herald an approaching stall are indelibly imbedded in my consciousness; I learn my lesson, which is the purpose of the whole thing. And I remember one evening at Danbury, when with a group of students I was discussing stalls with Gene. "What happens when you stall in a climbing turn?" I asked him. "Oh boy," was his answer. "Whoosh!" And with a sweeping gesture of his outstretched palm he made a picture of an airplane sliding helplessly and speedily off sideways. It didn't answer my question; it frightened me, and I carried that fear inside me to this day, when I found out, 3,000 feet—thank God—not 300 feet high.

We do stalls in a gliding turn, too. This is the type of stall that can occur when the pilot is coming in for his landing. It is no less dangerous, but it is far more difficult to do because the plane resists it even more. I could force 13-Delta into a climbing stall; I find I cannot get her to stall in a gliding turn at all. She justs shrugs, in a ladylike fashion, drops her nose and starts flying again. She loses altitude, of course, but she doesn't fall off sideways or do anything sudden at all. "What the hell," I finally say to Steve, "I just can't seem to make her do it. What's the flight inspector going to say to that?"

"He can't flunk you for not getting a stall out of her," Steve says. "After all, these airplanes were built to resist stalling as much as possible. Let's see what I can do." He takes the wheel, starts his glide, then turns. He lifts the nose a bit and 13-Delta starts her ladylike shrug again. At that instant he pulls the wheel sharply all the way back. Startled, the airplane rises as if to say: "Now look here . . . ," then, with a small shake, drops off over to one side and, muttering, picks up speed again. "That's about the best you can get," Steve says.

Once, flying along, Steve asks me to cut back my power and show him some slow flight. This is an art I feel I have long since mastered, flying along in the Danbury traffic pattern behind the far slower Aeroncas. I cut back to 1,700 r.p.m., trim up and we mush along at about 75 mph. "Not slow enough," says Steve. "Get her back more." I juggle throttle and trim control and get down to 65. "Come on," says Steve, "you can do better than that. Let's see you get down to 60. Watch your air speed and your altimeter; I don't want to see you lose any altitude."

At 65 I feel as though we were almost standing still; cutting back to 60 is an infinitely delicate maneuver. I ease back the wheel a hair more. We are mushing very nose-high now, and I can feel the lift fading from the wings. The air speed creeps back—64, 63, 62. We seem to hang on the very edge of nothingness; yet the ship still flies. I touch the trim tab again. The engine is ticking over so quietly that it fades from my consciousness; all my attention is riveted on keeping her flying. Sixty-two, 61, 60. And still she flies. The needle touches 60, 59, 58. "All right," says Steve. "Now give me a left turn."

A turn! Good Lord, here we are barely hanging in the air and the man wants a turn! I don't think it can be done, but I'll try. Gently, very gently, I ease the wheel over. The wings dip, hesitate, dip just a bit more. We start pivoting slowly over the sunlit sea. I neutralize the controls and watch this phenomenon in complete fascination. It is really beautiful. We are flying so slowly and quietly that the ship seems to be floating on a bubble of lift, a delicate, invisible, tenuous sphere of air that would break and dissolve at the slightest rough movement. Yet she flies, she flies! It is a miracle.

Straight and level again, just floating along, Steve asks for a stall. We are so close to one that I feel I need only ease the wheel back. I do, and for a moment we hang there, at the ragged edge of flying speed. "Now that," says Steve, "is just about the most dangerous thing you can do."

Startled, I ask him: "Why?"

"This airplane," he says, "is built so that it practically can't spin. But you're just begging for a spin here. You're right on the edge of a stall, but you're not stalling her. If this were an old-fashioned plane you'd be in a spin so fast you wouldn't know what was happening. You've either got to stall her or get some flying speed. Here, look. Let me show you."

One-three-Delta, not having stalled, is mushing along quietly again at little less than 60 miles per hour. Steve takes the wheel and eases it back. Once again, she hangs. He gives a slight kick to the rudder.

Whoosh! Sky, sea, and ground dissolve in a sudden blur. The left wing goes down; 13-Delta pivots on her nose. For an instant everything whirls. Then we are diving. Steve brings the wheel back, feeding her power at the same time.

"That wasn't a spin," he says. "We're not allowed to spin this plane, and anyway I'd really have to kick it to get it into one. But that's how a spin happens.

"You're sitting there, really hanging on your prop. Your lift is just about all gone. The airplane has to go somewhere.

"Leave it alone, and it will stall—not a clean stall, but it'll stall some way or other, drop the nose and try to get some flying speed. But if you make any kind of mistake, like kicking in a bit of rudder or putting your wheel over, she's going to go right over sideways. This kind of slow flight, of course, is not recommended, but it's something you've got to know how to do for your flight test. And you might as well also know what to do if you find yourself on the edge of a stall in these conditions. They may ask you for a stall in slow flight; if they do, give them one, quick and clean. Otherwise, if you ever find this happening to you, get that wheel forward and that power on, fast."

The next day, in another ship, he showed me what a real spin was like. We went up and duplicated the situation exactly. At the edge of the stall, he kicked left rudder.

Immediately, in bewildering dissolution, the world exploded before my eyes. Bits of sky flashed past the windscreen; then bits of ocean, then ragged streamers of land. Up and down were gone; there simply was nowhere that was up or down or even sideways. My stomach shot up to my throat, my eyes bulged, and I strained against the seat belt —looking straight at the whirling earth. It spun and spun in a dizzy blur; then suddenly it stopped spinning and, even more terrifying, I found myself hanging vertically, looking straight ahead at a brown field three thousand feet below. Then the engine sang, the field tilted sharply, sky appeared, my stomach sank back to its accustomed place and we were level again. Steve's voice was in my ear: "Now you do one."

Oh my God, I couldn't. Not for anything or anyone. I couldn't possibly let the world go crazy like that again. Fear churned in my bowels like ice water. I couldn't do it. But I had to.

Slowly I brought the nose up again, slowed the plane down.

We hung, in what seemed to me a terror-stricken silence, over the void. "Okay," said Steve, "kick left rudder."

I stabbed at the pedal with a tentative toe. The ship teetered. And then we went over.

This time I saw it and felt it: we went right over, almost stood on a wing, then slid and spun. Once again, earth, sea, and sky whirled. "Opposite rudder!" Steve shouted in my ear. "Wheel forward!" I shoved at the controls. And we came out of it, diving cleanly.

I did three more spins that day, and by the last one I actually knew what I was doing. I was still scared blue at every one of them, but I found, to my own surprise, that I was getting a thrill out of it too; an edge of excitement was eclipsing my fear as I counted those whirling turns—one, two, three; opposite rudder; out.

Okay, Gene, wherever you are, I said to myself as we flew home, *I really know what happens now!*

It is late afternoon of another beautiful day. I have been up in 13-Delta practicing steep 720-degree turns—twice around in a tight circle—around a pylon. Banked at nearly 60 degrees, engine roaring, wheel hard back to keep my altitude, I have been spinning, first to the left, then to the right, around and around a pond and a point at Chappaquiddick, across the inlet from Katama. Now I am sitting on the lawn outside the snack bar, relaxing, enjoying myself. Faintly I hear the Unicom speaking inside Steve's little office where his wife, Dorothy, is keeping watch on the airways and on the account books. Then, with a whistle of wings, 41-Papa, Steve's Apache, comes over the telephone wires and sits down for a landing. Steve turns the ship at the end of the runway, taxis back and helps out his passengers, two week-end Vineyarders who have come in from Providence. Full of bounce and energy, he comes striding over. "Come on," he says, "let's fly. I want to show you something."

Patient 13-Delta taxis out again, points into the gentle evening breeze, makes her short run, and soars. We rise to a thousand feet, turning over the darkling sea. Then Steve takes the controls. He hunches over them, having fun. He banks the plane steeply and we go roaring back over the field. "Chandelle!" he shouts gleefully and we swoop and soar and turn again. Downwind and parallel to the runway, he turns the ship back over to me. "Now," he says, "let's see you make a 180-spot landing."

This is the sort of thing he loves to do. He has sprung a surprise on me, and he knows it. Well, all right, if he wants to play I'll give it a whirl. I know what a 180-spot landing is in theory. You cut the throttle at the end of your downwind leg and, gliding in, try to hit a predetermined spot on the field. Since it involves a 180-degree turn, this takes some planning. "There's your spot right there," Steve adds. "See that bare patch on the runway just beyond the intersection? Let's see how close you can get to that."

We are a thousand feet up, just coming past the end of the runway. I look out the side, gauge my distance carefully, cut my power, and turn. The engine dies to a murmur; the wind whistles past our wings. Now I am on base leg. I turn again on final, pulling first flaps. We are coming down fast. I glance at the air speed, trim up a bit, slow down. We're getting low. Too low. "If I were you," says Steve,"I'd add some power."

I push the throttle; the engine comes to life; we rise and skim over the wires. "Here," says Steve, "let me show you."

He takes over, the engine blasts, we rise, turn and swing into our downwind leg again. "Now," says Steve, as we come opposite the end of the runway, "here's where you start your landing."

He cuts the power and at once starts turning. We sweep around in a great, soaring arc. He straightens for his base leg, squinting carefully at the field. "We're a bit high," he says, "so we S-turn a little to lose some altitude. We want to hit that field just right." We sweep out beyond our normal glide path, then turn back in. "Okay," he says, "that should just about do it." The wires flick past below. "Full flaps now." We balloon slightly and sink. We are just skimming the ground. "Now," he says, "there's our spot. We add a touch of power, so-o-o"—the engine murmur rises to a gentle song. "Now we hang her on her prop"—up comes the nose and we sail along three feet above the grass. "And here we are"—he cuts the power, 13-Delta sinks, touches lightly, flicking up a spurt of sand from the bare spot on the runway. Steve grins like a happy boy. "See?"

This is more fun than anything. I take off and try it again. This time I am much closer, but I add my power too late, hit with a bump, and bounce into the air again. The third time I begin to get the hang of it. It is absolutely exhilarating. I sail down the runway just over the grass, cut my power, and down she comes. I've talked about landing on a dime before; now I know what it feels like, and it feels great.

For the next few days I practice spot landings endlessly. I am getting better and better. Then, suddenly one day, I lose it completely.

I am up flying with Steve, getting ready for my final checkout. I cut my power, turn and glide. A quarter of a mile from the runway I am all out of altitude. Shamefacedly, I add power, climb and turn again.

This time I am a little closer, but not much. "I can't understand it," I tell Steve as we climb out and away for another try. "I'm undershooting every time."

"Well," he says, "of course you've got more weight in the airplane with me in it. Try starting your turn a little sooner." But I can see that he, too, is puzzled.

Finally he tells me to take the ship over to the County field, the big Martha's Vineyard airport with its wide, paved runways and long approaches and try it there. I shoot one landing, come in way high, touch halfway down the field. By this time I'm beginning to sweat. Steve gives me a demonstration, then turns it back to me. I have been watching his every move, and I try to duplicate his landing exactly—same height, same place to start the glide, same turn, everything.

And now we find the answer. I turn, and 13-Delta slides off and down and away. "Look at your altimeter," says Steve. "Why, you're slipping that turn like crazy. You'll have half your altitude gone before you ever get on final. And altitude is your insurance. You've got to have it. If you have too much, you can always slip it away at the end—but if you don't have enough, well, I can tell right now that you're not going to make it, not by a long shot."

And of course I don't. But the next time around I watch my turn like a schoolboy doing a math problem. Instead of figures, I have degrees of bank, touches of rudder, jockeying of wheel. If I add them all right, the problem will prove out with a landing where I want it to be. Concentrating as I have been before on only one aspect of the problem—height versus distance—I have neglected the fundamentals. Carelessness in turns is perhaps the most common fault in any pilot, and I have been either slipping mine—in which case I have lost altitude, sliding down sidewise as I turned—or skidding them, in which case I have lost speed, slewing around sidewise like a car sliding on a slippery road. Now I try, and the problem proves out. I don't even have to add power—13-Delta skims over the fence and kisses down right on the runway numbers.

The next day Steve takes me up for my final checkout. He rides me hard. We do all the stalls; we do slow flight; we do steep turns around a pylon, we do 180-spot landings and we do a simulated forced landing, spiraling down with no power at all. We climb and stall and turn and land for one hour and fifteen minutes in the cloudless sky over the Vineyard and when it is all over and I have filled out my logbook with a hand that trembles from nervous exhaustion, Steve adds his own statement in handwriting that sprawls boldly across the pages:

"All Phases of Flight Test Procedures Release No. 420 Satisfactorily Completed and Applicant Recommended for a Private Pilot Certificate."

An hour later, just as we are sitting down to dinner, he calls me on the phone. "Day after tomorrow," he says.

"Eight-thirty in the morning at Norwood. I just made the appointment. That's the best time of day; you'll catch the inspector before he gets grumpy with a lot of bum flying. You're ready, and you'll be okay."

Seven A.M. I hear the click of the radio as it turns itself on and realize that the day has inescapably begun. I get up, put the coffee on, and walk out to the pool for my early-morning swim. The air is fresh and cool. The sun has not yet come up over the low haze of ground fog that lies over the Vineyard; fog that will be dissipated, I know, within a half-hour. Above, the sky is a clear blue. It is a day like all the other Vineyard days except for one thing: this is the day when I am taking my Private Pilot flight test.

It is odd how, at the age of forty-five, a man can still dread an examination. I had a schoolboy's nervous fear of the written test, but this is different. This is the payoff on a whole year of learning to fly. I did not learn quickly; I did not want to, and other circumstances combined with making my training period far longer than is usual. Mostly it was the fact that I could not learn consecutively at any one school. Beginning with Steve Gentle at Katama, I had gone on to three different schools, half a dozen instructors and, finally, Steve Gentle again. It seems a long, long period of learning, and now I suddenly realize that I have not only an unbecoming dread of my flight test but also a feeling of sadness that it is all coming to an end.

How many week ends had there been when, rising, my first look had been to the sky to check the weather and the wind! How many hours of sitting with an instructor at my side! My total time in learning was about fifty-six hours, including checking out again for solo flight, after Katama, at Danbury and Teterboro, with a couple of hours at Bridgeport thrown in. This, I knew, was about average, but stretched out over week ends through autumn, winter, spring, and summer, it seemed a long, long time.

And often—too often—during that learning period I had felt that at least half my job in learning to fly was pleasing my instructor. This was a mistake, I knew, but gradually I had come to accept it as something I had to do. Every instructor has his own way of doing things and naturally he teaches in accordance with that way. And to a student, any student, an instructor is for a long time a sort of demigod, a man he must perforce look up to, the teacher who has mastered this mysterious art of flying. I had often discussed this with Bill Strohmeier and had pointed out to him how few were the instructors who seemed to be really aware of this immense responsibility they held, how often the teacher totally disregarded the student's early hesitations, uncertainties, and fears. And how often one instructor would contradict another's teaching—and here I remember an incident now which makes me smile.

It happened when I was being checked out for solo flight at one of the three schools I had attended. I have spent several hours getting used to this field, where conditions were totally different from any I had known before, and to the man who flew with me. One of the things he taught me was to round my corners in the traffic pattern—and now, just before my check flight with the chief pilot, he reminded me once more: "Now remember, Perc, at 400 feet start a wide, gentle turn all the way around into your downwind leg. That'll give you plenty of chance to look around and check for other traffic. Hold that turn all the way around, nice and smooth. . . ."

We landed, and the chief pilot climbed in. I took off, and at 400 feet I started a wide, gentle, climbing turn. I held it all the way around, flew my downwind leg, turned into base and final, and landed as gently as a thistle. The chief pilot said never a word until we had taxied back to the hangar. Then he turned to me. "That was fine, Percy," he said. "I like the way you flew. There's only one thing—I want you to learn to square off your turns on the pattern."

Well, what are you going to do? Flying solo, I decided that I preferred to square my pattern, and I have flown thus ever since. But what about this man, the inspector, with whom I am to fly today? Is he a squarer-offer, or is he a wide-turn man? Whatever his philosophy of flying the pattern, he is the demigod of all demigods—he is the final authority, the last of a long line. If I don't please him—boy, that will be it; I'll have to start all over again.

So here I am, after all this time, still suffering from instructoritis. It is an insidious and dreadful affliction, and all of a sudden it seems a damned silly one to me as well. In the long run, a pilot who has been well trained can fly only one way—his own; and please only one person—himself. He is the man in the left seat, the pilot in command, and it is up to him to get his ship safely home. To hell with it! I will fly as best I know how, and if that isn't good enough by this time (inwardly I am sure it will be), well, then I deserve to get sent home to learn some more.

Half an hour later 13-Delta and I are aloft and winging our way across the Vineyard toward Norwood and our climatic occasion.

It is a rarely wonderful day for a flight test. Though the sky over the Vineyard is a clear blue, a high overcast hangs over the mainland—a great roof of slate-gray cloud that stretches ahead and on both sides along the coast as far as the eye can see. And I can see for miles and miles. Beneath that roof of cloud I can see, to my left as I cross Buzzard's Bay, the rocky point off Newport and the smoke of Providence beyond. On my right is the length of Cape Cod, right out to its curving tip. Just a few miles away in that direction is the vast expanse of Otis Air Force Base, a tremendous field. Ahead are the radio towers at Marion, a whole forest of them, a beautiful landmark. And far, far ahead is the bulk of Blue Hill, the ridge that marks Norwood. And we fly toward it in air that is as smooth, as steady as an invisible carpet. My ship is squared away on course and she can fly herself; there is not the slightest tremor to upset her. It is a stillness so all-encompassing, so beautiful that it seems devoid of any sound; the peaceful drone of my engine seems only to accentuate it, to increase the feeling of absolute perfection of flight.

And so we fly on in the peace and the stillness and presently Norwood comes up below. I circle the field at 2,000 feet and check the wind tee. It points straight up the north runway, a sheet of gray macadam on the brown earth as wide and inviting as a ballroom floor. I cut power, we turn and glide down slowly into the downwind leg. I announce myself on the Unicom, check for other traffic and find none.

We soar across the fence at 65 miles per hour, touch lightly just beyond the markers, and roll. If only, I think to myself, Mr. Unknown Flight Inspector could have seen that landing—well, at least it makes me feel good.

The silence of the upper air hangs over Norwood, too, as I climb out. I don't see a soul. I climb the short path to the operations office, noticing the signs along the way: HAVE YOU CLOSED OUT YOUR FLIGHT PLAN? and other reminders of what a pilot should do when he has come home. And then, with a strangely empty feeling, I open the door, walk in and announce myself: "Percy Knauth from Edgartown; I'm here for a Private Pilot's flight test."

From that moment on, I am caught up in a round of events in which time and all outside things cease for me. I meet my man, a pleasant, taciturn, tweedy, pipe-smoking type, and he invites me to sit down. He studies my logbook, paging through it with utmost care. He looks over my student pilot's certificate, checks the report card of my written examination. "All right," he says at last, "let's talk a bit about flying."

This is the quintessence of every hangar or snackroom discussion I have ever had on this subject which has become so dear to my heart. It is Phase One of the flight test—the Oral Operational Examination.

This man doesn't just ask me questions; he involves me, as one pilot to another, in a discussion of whys and wherefores. Why is it required, for example, to have the three documents that identify an airplane always on hand in the ship—the registration, airworthiness certificate, and equipment documents? He asks for them, and I give them to him, and together we study them. When was 13-Delta last inspected? Here it is—only two months ago. She had a good top overhaul—valves ground, a couple of new ones installed; new plugs; ignition overhaul—the works, plus a couple of new tires. "You've got a good airplane there," the inspector says approvingly.

Then we get into range of performance of the Tri-Pacer. We pull out the owner's manual and study it, checking figures. He asks me questions about my fuel supply; I answer them, remembering that there is always one gallon of unusable fuel in the bottom of each tank, and that the flight should be calculated so that there are 45 minutes of fuel left when the airplane arrives at its destination.

Well, now, suppose we were leaving for a flight right now? Suppose I had one passenger and a hundred pounds of baggage? How would I stow it?

Immediately, a conversation at the snack bar of a few evenings before comes to mind. A Tri-Pacer has a listed baggage capacity of a hundred pounds. That seems clear enough—but is it the whole story? It is also built to carry four people along with those hundred pounds. But if you have only two people, where would you put all the baggage?

It's an interesting question, and here in the flight inspector's office we once again discuss it thoroughly. In a car you would probably put all the baggage in the trunk. But in an airplane?

"Well," I say, "it's a question of balancing your plane. If you've got only two people in a four-place airplane, I'd put at least some other baggage in the back seat so as to balance it better."

"All right," he says, "let's take a look at your weight and balance form."

I know what he means; in that snack bar discussion the other evening we went over it thoroughly. I pull it out, and we look at the figures. With four people, the baggage goes where space is provided for it—in the baggage compartment. With fewer people, it can be worked out differently—so much in the baggage compartment, so much in the back seat. It doesn't *have* to be—the airplane will fly even if it is sadly unbalanced—but it *should* be, because if the load is properly distributed, the airplane will always be in proper trim.

We spend nearly forty minutes on this sort of thing. Then the inspector gives me back my documents. "All right," he says, "now let's get ready for a little trip. Do you have an Albany Sectional Chart with you?"

I have. He tells me he wants me to lay out a course for Westfield, an airport northwest of Springfield, Mass. "Get yourself some good check points," he says, "but don't make them too far apart. We're not going to go all the way. I would suggest that check points every four or five miles would be about right. Work it out and come back here in—let's see"—he glances at his watch—"ten minutes."

I go outside and bend over my charts. The course to Westfield is just about due west from Norwood. I mark my check points with care. I can't get them quite as close as he suggested; I get a couple three miles apart, the rest five or six. But they are all good ones, and I hope for the best. I go back in, present him with my figures—my course, based on predicted winds aloft, my estimated time of arrival, my alternate airport, my check points. He studies everything gravely for long minutes, then reaches for his hat. "All right," he says, "let's go."

He does not ask me for a line check of the airplane, but I know that one is required. I know, too, the kind of line check that I like to make. I go over 13-Delta from propeller to rudder, checking her oil, her spark plug connections, running my hand along her wings, checking aileron, flap, rudder, and elevator hinges. The inspector, meanwhile, stands there in the chilly air, smoking his pipe, saying nothing. It is I who says at last: "Okay, we're ready." And we climb in, run up the engine, announce our departure and go.

With virtually no wind, the take-off is smooth and we climb out strongly. The inspector, sitting erect beside me, hat on his head, his cold pipe in his mouth, allows himself one comment. "You've got a good day for a flight test," he says. "The air is nice and smooth. You're lucky. Now let's get on our course and start for Westfield. I would suggest we fly at 2,500 feet."

I circle the field once as we climb to altitude, then get squared away. My map is spread out on my knees and I follow it with a careful finger. I have figured a ground speed of about 80 miles per hour, considering the prevailing winds. This is perhaps a little slow; my first check point, the town of Westwood on my right, comes up very quickly. And the others follow right along: a crossing of road and railroad three miles further on, and there is the town of Millis off to my left, and two towns together beyond. "Now," the voice of the inspector cuts in, "take me to Marlboro."

For an instant, I am stricken. Marlboro! I haven't the

faintest notion of where it is. North, west, or south? I begin searching in widening circles on the map. "It's north of here," the inspector says helpfully. Ah, there it is! I make a rapid check with my plotter—yes, it's northwest, about fifteen miles away. Now let's see—that means a 45-degree turn, approximately, to the right; my present heading is 269 degrees; that means a new heading of about—269 plus 45—314 degrees. It will have to serve for the moment. I turn the plane, get her on course again, then go back to my map for more accurate figuring. A quick pencil line from just beyond Millis to Marlboro defines my new course; it checks out pretty well; now what are my check points for this leg? Ah! There's a big town about eight miles away—Framingham—I can't possibly miss that. And a lake right beyond it, and right at the tip of that lake is the airport of Marlboro. We're back in business; I settle back in my seat and wait for Framingham to appear.

But it doesn't appear. The minutes tick by. Five minutes have passed; I should see Framingham now. I scan the horizon. Off to my right is a big smudge of smoke, but ahead is nothing but a scattering of little towns and villages in a wooded countryside. There's another big town on my left, a really big one; could that be Worcester; am I so far beyond where I thought I was? Slowly I get a panicky feeling: I am lost.

Beside me, the inspector has fired up his pipe. He sits there, gazing pensively out of the window. No help, no hope from him. He doesn't say a word, he doesn't indicate even by a lifted eyebrow whether everything is all right or all wrong.

I scan the horizon again. Now I see a lake opening up ahead, a big one. Aha! I am right, after all! But no. Framingham should be at the near end of it, and at the near end of this lake there is nothing at all.

Back to the map. Where is there a big lake in this general vicinity with no big town at the head of it? It is a long lake, too; it stretches square across my course, from southwest to northeast. And here is one that answers its description in every detail: Wachusett Reservoir.

Now I have found myself. I am, at this point, about ten miles west of Marlboro. I know it with certainty. But I can be a sphinx too. Without a word or gesture I turn the plane due east. And in a few minutes, sure enough, there is the town, there is the lake, and now I see the airport nestled in the low hills, a grass field with a few airplanes parked around. I bank the plane and point it out. The inspector nods. "All right," he says. "Let's head back to Holliston."

We turn south and fly in silence. Then he takes his pipe out of his mouth and points ahead with it. "See that pond out there?" he asks.

"Yes," I reply.

"All right," he says. "Now do you see that old sunken tree at the head of the pond?"

I squint out at the rapidly approaching small body of water. There's something there that looks like an old tree. "Yes," I reply.

"Go down to 400 feet," the inspector says, "and give me a steep 720 around that tree."

I have a lot of altitude to lose in a rather short time. I cut my power slightly, put on carburetor heat, and 13-Delta slants steeply downward. At 400 feet I level off and head for the tree. I check my wind direction—okay. I set my right wingtip on that tree, bank over and around we go.

Steep 720s have a feeling all their own. The motor roars; the wheel is held hard back, the world spins. The main thing is, keep your wingtip on the pylon; shallow it out on the upwind side, steepen it on the downwind side. Around and around we go. Midway in the second turn the inspector says: "Let me take it a minute. I want a *really* steep turn."

He swings 13-Delta over until she seems to be standing on her wing. We spin like a dervish. Pipe in mouth, he holds her there with one hand as casually as though we were 5,000 feet high going straight and level. "You want to get in closer," he says to me over the engine's roar. "Get in close and hold her right on there. See? Now you take over."

I hold that turn as though my life depended on it. Suddenly it is almost exhilarating: here we are, banked at 60 degrees with the world spinning by. There goes the pond—once—and again. "All right," says the inspector. "Straighten out and climb back to 2,500 feet."

For the next few minutes I am busy trying to remember where I am and where I should be heading, and we fly in a silence which is by now familiar. Lacking specific instructions, I work 13-Delta up to altitude again in a wide circle and head for our original course, planning to intercept it near Holliston. But the voice cuts in again. "Tune in the Boston omni," the inspector says quietly. "Get on a radial, and take us there."

Boston's frequency is 117.7; I switch on the set and tune it in. I am almost to Holliston now; by the map my heading should be about 30 degrees north of east, so I turn my course selector dial to 60 degrees. The needle comes alive; I tune in carefully and wind up with the needle centered at 78 degrees. Well, that allows for 14 degrees of magnetic variation which I had forgotten. We head toward Boston.

And now I really get an exercise in omni navigation. The inspector turns me off my heading, asks me to intercept a different radial, and fly that. I do so, concentrating hard. Remember the railroad track; remember to intercept it at an angle; remember the thread of the radial held lightly between the fingers. We intercept and fly toward Boston again, closing in on the narrowing radio signal. And then he turns me completely around and has me fly outbound, on the same radial, intercepting it again.

This is really tricky; this is the one time when the system works backward; when instead of flying into the needle one must fly away from it to get onto the radial and stay there. My course is about 260, and I am flying away from the station; but I am not on a radial FROM, I am still on a radial TO. So, in a sense I am flying backward, and I get the needle centered again by flying left when it shows right, right when it shows left until I am on course. Finally he turns me around, heads me back toward Boston again, and then requests that I take him back to Norwood.

This proves not too difficult; I can project my heading to Boston, draw a quick line on the chart and figure out approximately where I am in this ocean of cloud-hung air. And, thank God, there is Blue Hill, looming up in the distance. I reach for the microphone as we approach the field, get my clearance to land, and come in. Still not a word from the inspector. He lets me taxi right up to the intersection, then tells me to turn into the taxiway and head back to the end of the runway. "Now," he says, "give me a short-field take-off."

For the first time, I have a moment to realize that Phase Three of my test, the cross-country flight, is for better or worse over and done with. The short-field take-off definitely belongs in Phase Two, Basic Piloting Technique, which we are obviously beginning on now. The principle is a simple one: faced with a short field, the pilot makes the most of it by starting his take-off run at right angles to the runway available, turning the plane as he guns the engine so that, when he straightens out on the runway itself, he will already have some speed. On a grass field he might have considerable speed because he can skid the airplane around. Norwood, however, is macadam; this is going to be a bit tricky.

I swing 13-Delta out onto the taxiway, turn and point her toward the runway. I hold the brake with one hand, the throttle with the other and steer with my feet on the rudder pedals. Then I let her go.

I don't know how fast we are going when we start to turn onto the runway, but it is certainly fast. She tilts alarmingly, and the tires squeal. I haul on the brake, ease up on the turn. Now we are heading almost straight down for our take-off run. I let go the brake, grab the wheel and gun her. We slew around, straighten and away we go. We're up to sixty before I know it, and I ease her into the air. The inspector points up with his pipestem. "Twenty-five hundred feet," he says.

Up there we do our stalls. We do them right and left, power on and power off. He has me slow-fly her, and then come the stalls in a gliding turn. Try as I may, I can't get 13-Delta to do more than give me her ladylike shrug. "All right," says the inspector at last. "Take it down and give me a spot landing just beyond the numbers."

At 800 feet I swing into the pattern on the downwind leg, cut my power as I pass the end of the runway and make my turn. No slips this time; this has got to be right. But as I turn into my final approach I see I have a lot of altitude left. I remember one requirement of Phase Two: slips, and a slip to a landing. I've got to slip anyway, and maybe I can kill two birds with one stone. I give her left wheel and right rudder and we slide off to the left in a long slant, losing excess altitude fast. Then I straighten out and head for the numbers. Here they come; add a touch of power; hang her on her prop—now! I cut the power; I am a bit high; we bounce slightly, then touch down. The inspector waves with his pipestem. "One more," he says. "Give me a little more power on this one. Hang her up there and let her come in."

I pour on the power and we go around again. This time we touch just past the numbers. Once more around; he wants a wheel landing now. I am flying with such complete concentration that I could probably give him anything without even knowing it—we come in fast, nose-high, greasing it on. Once again we go around; this time he wants a landing in a full stall, as short as I can get it. With full flaps, we come in over the fence and I hold her off until I can hold her no more. Down she comes, and he waves me into the taxi strip. "All right," he says, "that's all."

In total silence I park the plane, in silence we clamber out. Now, for the first time, I notice how my knees are shaking, how wet I am with sweat. I see young Stevie Gentle, Steve senior's son, standing there on the taxi strip and suddenly remember what I had completely forgotten: that he has brought a Tri-Pacer up for a 50-hour inspection and that I am to fly him home. He gives me a questioning look; I give him an answering shrug. I honestly don't know.

Back in the office, when I sit down, I find I am trembling like a leaf. I have been flying for an hour and fifteen minutes; I feel as though I had been flying all day. The inspector removes his hat, sits down, lights his pipe, and reaches for a form. For a while he writes in silence. Then he extends his hand. "Your logbook, please," he says.

Wordlessly, I hand it over. He leafs all through it again. Then, deliberately, he proceeds to inscribe our flight in it. I cannot see what he is writing. I have a sinking feeling in the pit of my stomach. I got lost. I didn't make my 720s steep enough. I flubbed my first 180-spot landing. I . . .

Without a word, he hands my logbook back to me. I read, transfixed: "9013-D; Norwood, local; 1 hr. 15 mins. Private Pilot Practical Test Approved Certificate ASEL issued," and his signature. And now, for the first time, he smiles. "Congratulations," he says, extending his hand. "I guess you know what you did wrong. I think you can take care of yourself. If you'll stop at the desk on your way out, my secretary will give you your temporary certificate; the permanent one should arrive in the mail in about two weeks."

At the desk, as I pick up my certificate, I see what "ASEL" means. There it is written, for all to see: PERCY KNAUTH —PRIVATE PILOT—AIRPLANE SINGLE-ENGINE LAND."

When I get outside, I suddenly find myself running. I run down the steps and to the airplane, waving my new license like a boy. Stevie pounds me on the back, I climb in, taxi out and we take off. All the way home I am talking, and when we get to Katama I give him one of the worst landings I have ever made. I come in high; I pull on flaps; I try to force her down and we hit with a bang. "Oops!" I say. "I'm sorry!" "Go to hell," says he. "Who ever told you you could fly?"

Well, now I know I can. But as I taxi to the hangar I know that it will never be the same again. Here, on Katama's worn brown sod, I made my first flight; here I soloed; here I groundlooped the Super Cub; here I put the final polish on a year of learning, and here I departed this morning, to be or not to be Private Pilot Knauth. And here I have returned, and bounced my first landing as a licensed pilot. It is as though that worn, that soft, that welcoming brown earth had told me: all right, boy, you're on your own now. Don't expect anything more from me. I'm not going to reach up and pull you down out of the sky. You can come back here any time, but you must know now that from here on in, you can set your wheels down anywhere, all on your own.